Secrets of the
Old One

Secrets of the Old One

Einstein, 1905

JEREMY BERNSTEIN

C

COPERNICUS BOOKS

AN IMPRINT OF SPRINGER SCIENCE+BUSINESS MEDIA

© 2006 Springer Science+Business Media, Inc.

Published in the United States by Copernicus Books,
an imprint of Springer.
Springer in a part of Springer Science+Business Media

springer.com

Library of Congress Control Number: 2005926763

9 8 7 6 5 4 3 2

ISBN-10: 0-387-26005-6   e-ISBN 0-387-25900-7
ISBN-13: 978-0387-26005-1

Contents

Acknowledgments

Writing about Einstein and his work is a daunting task in which I have been helped or encouraged by a number of people whom I acknowledge here. There is the usual disclaimer that they are not responsible for any mistakes that may have crept into the text. On the physics side I would like to thank Elihu Abrahams, Freeman Dyson, Michael Fowler, Murray Gell-Mann, Owen Gingerich, Sheldon Glashow, Jim Hartle, Gerald Holton, Dave Jackson, Tom Jacobson, Michel Janssen, Eugen Merzbacher, Arthur Miller, John Rigden, Wolfgang Rindler, Andre Ruckenstein, Engelbert Schucking, Robert Schulman, John Stachel, and Roger Stuewer. On the book side I am very grateful to Harry Blom for keeping the faith and to Chris Coughlin for his able help and to Barbara Chernow and her staff for the production of the book. To each I offer the Viennese wedding toast, "May the parents of your children be rich."

Introduction:
Einstein's Miracle Year

Everything should be made as simple as possible, but not simpler.

—Albert Einstein

eginning on March 18th,1905, and ending on June 30th, at roughly eight week intervals, the leading German physics journal *Annalen der Physik* received, in its editorial offices in Berlin, three handwritten manuscripts. Written by a patent examiner in Bern, Albert Einstein, they would in their totality define physics for the next century and beyond. A fourth briefer paper—really an addendum to the third—was received by the *Annalen* on the 27th of September. It contains the one formula, $E = mc^2$, that everyone associates with Einstein. These papers, which are the subject matter of this book, are remarkable in many ways. First, there is the manifestation of creative scientific genius. Nothing like this had been seen in science since the year 1666–the *annus mirabilis* (the miracle year) —of classical science.[1] That year, 23-year-old Isaac Newton, who had sought refuge in his mother's house in Lincolnshire from an epidemic of plague that was devastating Cambridge, created the basis for physics that endured for the next two-and-half centuries. Second, there is the style. Einstein's papers contain very few references to the contemporary literature. They only rarely refer to each other, something that, as I will explain later, would in at least one significant instance have helped readers to comprehend them. This paucity of discussion of contemporary literature is one of the reasons why the papers appear so fresh. There were other very important papers of the period, some having to do with the same general subject matter, but they seem dated. One has to peel off the parts that are still valid from the parts that are not. Although vast progress has been made in our understanding of the physical world in the last century, nothing of any importance in Einstein's papers is wrong. One can teach the theory of relativity from the third paper, and one can also teach the implications of the quantum nature of light from the first. In all the papers, the writing is elegant and economical. We feel that we are in the sure hands of a master —a master who was, at the time just twenty-six.

It is not my intention to present a biography of Einstein. There are innumerable biographies, and the number is growing. But I want to describe the years leading up to 1905 to make clear the context in which the papers were written. Einstein was born on March 15, 1879, in the southern German city of Ulm at the foot of the Swabian Alps. His parents, Hermann and Pauline Koch Einstein. were Jewish, although not very practicing. There is no trace in Einstein's genealogy of anyone with scientific accomplishments. This certainly had something to do with the professional restrictions that were placed on Jews in the ghettoes. In fact,

---

[1] Historians of science note that the term, which was originally used by the poet John Dryden to describe the English victory over the Dutch in 1666, better designates the period in Newton's life from 1664 to 1666.

it was only in 1871 that Jews were recognized as full citizens of Germany. As a child, Einstein was very slow to speak. It worried his parents. In 1947, Einstein was persuaded by the philosopher Paul Schilpp to write a sort of autobiography, something that Einstein referred to as writing his own obituary. Actually he died in 1955. It is published as the introduction to an extraordinary collection of essays written in his honor. Most of his autobiography deals with his science, but a little of it describes his early life. At the age of four or five his father gave him a compass whose behavior made a lasting impression. He writes, "That this needle behaved in such a determined way did not at all fit into the nature of events which could find a place in the unconscious world of concepts (effect connected with direct 'touch.') I can still remember — or at least I believe I can remember — that this experience made a deep and lasting impression upon me. Something deeply hidden had to be behind things." Some years later, in his early teens, Einstein discovered Euclidean geometry. In his autobiography Einstein tells how he found for himself a proof of the Pythagorean theorem which relates the sides of a triangle with a right 90° angle. This theorem will be one of our main mathematical tools and, later in the book, I present my reconstruction of Einstein's proof.

Einstein's father was a not very successful businessman specializing in various electrical equipment enterprises. When Einstein was one year old, the family moved to Munich so that his father could set up a business with his younger brother. So, when Einstein was ready to go to school, he entered a so-called "Gymnasium"—in this case the Luitpold Gymnasium. In this school, which was a state-supported Catholic school, there was essentially a military discipline. The students wore uniforms and were drilled. Einstein thoroughly disliked the place. It strengthened the pacifist instincts he had had since early childhood and which he only abandoned in the 1930s with the rise of Hitler. It is sometimes said that he was a poor student, but he was, both in high school, and later when he entered the *Eidgenössische Technische Hochschule*—the Swiss Federal Institute of Technology in Zurich—which Einstein came to refer to as the "Poly"—a good student. He was never at the top of his class but he was always above average.

Einstein's problems at the Gymnasium, and the Poly could be attributed to what his teachers perceived as an attitude. He never had much respect for authority, especially if it was associated with a manifest lack of competence. It reached such a point at the Gymnasium that, by mutual consent, Einstein withdrew in December of 1894. By this time, his family had moved to Italy, where his father started another ultimately unsuccessful business. Einstein had been left to live with relatives in Munich, but in 1895 he joined his family in Italy where he spent what he remembered as a delightful six months. Part of the time he studied for the entrance examination to the Poly. He took it at age sixteen-and-a-half and did well in the scientific parts but not very well in the rest which dealt with languages. He was advised to take an additional year of study. For this purpose, he chose a progressive school in Aarau, Switzerland. By this time he had decided to give up his German citizenship, which really meant giving up his citizenship in the state of Württemberg, which was done for a payment of three German marks. He remained stateless until 1901, when he became a Swiss citizen. Like all Swiss men, this meant that he was obligated to serve in the army. He was exempted because of flat feet.

In 1896, he passed the entrance examination and spent the next four years at the Poly. In his autobiography, however, he wrote that he could have received a better education, especially in mathematics, than he did, as there were very good mathematicians there whose courses he was not interested in. He also decided the teaching of physics was inadequate, so he spent most of his time teaching himself. He complained, for example, that the electromagnetic theory of the Scottish physicist James Clerk Maxwell, the greatest advance in physics since Newton and which was then some twenty five years old, was not being taught. He had to learn it on his own. Einstein's professors were aware that he was not attending all his classes, and they did not appreciate his attitude. Nonetheless, his grades were quite good because he studied from the meticulous notes of his friend, Marcel Grossman, who later became a mathematician with whom Einstein collaborated. But when he graduated, he was not asked to stay on as an instructor or laboratory assistant, something that several of his fellow students were invited to do. His teachers did not want him around. He then tried unsuccessfully to find employment in several physics institutions in a variety of European countries. This was certainly due in part to anti-Semitism, but it was also the result of what were very likely not very enthusiastic letters of recommendation.

Einstein began a two-year period of odd tutoring jobs. One wonders what would have happened if Marcel Grossman's father had not helped him to get a job in 1902 as a patent examiner at the Swiss Federal Patent Office in Bern. He became a "technical expert third class," with an annual salary of 3,500 Swiss francs (see Figure I.1). I have read different accounts of how much time his job left him for doing physics. One thing is certain. It was a serious job which he took seriously. A few of his patent assessments are still extant. They are thorough and sometimes sharply negative. Einstein may have, especially in his later years, looked like a benign presence, but he had a very cutting tongue that also got him in trouble. He had no time to actually carry out calculations during patent office working hours, but that nothing could stop him from

Figure I.1. Einstein at the patent office. (Courtesy AIP Niels Bohr Library)

thinking about physics. One reasons, why he did not have a better knowledge of the contemporary physics literature was that the university library in Bern was closed when Einstein was free on nights and weekends. I think it is also true that he did not much care and did not want to waste his time reading about physics that he was quite sure was wrong.

With his new job he was able to get married. While at the Poly he had met a fellow student, a somewhat older Serbian woman named Mileva Marić. The Poly was one of the few places in Europe where a woman could study science. Their relationship started as a school friendship, but by 1898, they were considering marriage. Einstein's mother was vehemently opposed. By the end of 1901, Mileva became pregnant and gave birth in Hungary to a daughter we only know by the nickname "Lieserl." Einstein never saw his daughter, and no one knows what happened to her. In any event, in 1902, Mileva and Einstein were married. In 1904, they had the first of their two sons, Hans Albert. The second, Eduard, was born six years later. The marriage ultimately ended in a painful divorce. Einstein gave, as part of the settlement, the proceeds of the Nobel Prize which he had won in 1921.[2]

This is the context in which the papers were written. I cannot imagine where he found the time. He had a full-time job, family responsibilities, and a social life. He played music—the violin—and had friends with whom he spent time. When could he work on his papers? Each of them has scores of equations. He must have been able to calculate with incredible speed and precision. To add to everything else, he wrote them out by hand for submission to the journal. Further, he was writing a doctoral thesis, published the next year, also written by hand.

Now let me explain this book. There are four chapters and an epilogue. The first is an account of the relevant physics history up to 1905, especially as it deals with electromagnetism and Newtonian mechanics. Other chapters recount the corresponding history for the subject matter at hand. The second chapter is an account of Einstein's papers on the

---

[2]He actually collected the Prize in 1922. The divorce was in 1919, after which he married his cousin Elsa Löwenthal Einstein.

theory of relativity, with additional relevant prehistory, and a sketch of what happened to the theory after 1905. The third chapter deals with what is known as "Brownian movement," that is, the random motion of microscopic particles suspended in liquids. This development, and the experiments it led to, persuaded most of the skeptics—and there were some important ones—that atoms existed as real physical objects and not as mathematical abstractions. The last chapter deals with the quantum. It was the first paper of the series chronologically; the relativity papers were the last. This first paper was the only one Einstein thought truly revolutionary. I will explain the reasons. The reader may be surprised that this chapter begins with the history of the steam engine. You will see why. It is important that I make clear my overall objective. I want to explain all of this using mathematics no more difficult than that taught in high school—simple geometry and algebra. This does not mean that I skimp on the ideas. I think that they are all there, as simple as I can make them—but no simpler. Before turning to the first chapter, let me explain briefly how I got into all this. It will also enable me to introduce you to someone you will meet from time to time in the book.

In the fall of 1947, I entered Harvard University as a freshman, where I discovered there was a science requirement. If you were not a prospective science major, which I was not, you had to take a Natural Science course in the then rather newly created General Education program. I took what was reputed to be the easiest one—Natural Sciences 3—which was taught by the late I. Bernard Cohen, a historian of science and a Newton expert. That is how I first learned something about Newton. Toward the end of the first semester, Cohen touched a little on Einstein's physics and a bit about his life. Einstein was then at the Institute for Advanced Study in Princeton. I learned that as people, Einstein and Newton had almost nothing in common. Newton was austere and virginal and spent at least as much time on biblical dating and alchemy as he did on what we would call science. There is only one recorded instance where he was heard to laugh. Einstein was bohemian, much interested in women, and loved to laugh. When he heard a good Jewish joke it was said that he had the laugh of a barking seal. Both men were

Figure I.2. Philipp Frank. (AIP Emilio Segré Archives)

in their ways profoundly religious. Both men were, and are, to historians and biographers and to me, endlessly interesting.

Although I understood relatively little of the science, it took hold of me, and I decided to learn more about it. Cohen told me that a successor course was being taught that spring and that I could, if I wanted, take both simultaneously. He also said it would be taught by a man named Philipp Frank (see Figure I.2). Frank, he added, had known Einstein for decades. Indeed, he had succeeded Einstein at the German University in Prague when Einstein left in 1912 to return to Switzerland, and he had just published a biography of Einstein, *Einstein, His Life and Times*. It sounded perfect.

The class met once a week, on Wednesday afternoons as I recall, in the large lecture hall in the Jefferson Laboratories. There were perhaps fifty students. Professor Frank turned out to be a shortish man with something of a limp acquired in an accident in his native Vienna, where he had been born in 1884. What hair remained was distributed around the side of his head in wisps. He had, I thought, the face of a very intelligent basset hound. His accent was somewhat difficult to place. I used to say that the languages he knew—God knows how many—were piled one on top of each other like the cities of Troy, with shards belonging to one popping through to the others. On one notable occasion in response to a question from a student, he wrote on the black board a quotation in Persian, a language he later told me, he had learned in night school in Vienna. He would lecture for about an hour and then announce that he would now make a "certain interval." After the interval, you could return to ask questions. Sometimes he would give an answer that he said could be understood "if you knew a little of mathematics." The only mathematics I knew was what I had learned in high school—a smattering of algebra, trigonometry, and Euclidean geometry. That is all you needed to know for his course. I decided to learn "a little of mathematics" and ended up majoring in it.

I think that the reason Professor Frank could explain things so clearly and simply is because he understood them so well. He had taken his PhD in physics in 1906 under the direction of Ludwig Boltzmann, about whom we will hear later. Professor Frank understood the importance of Einstein's physics from the beginning, and was soon in contact with him. He made significant contributions to the development of relativity. We owe to Professor Frank the term "Galilean relativity." We will soon examine Galileo's relativity and learn what the term means. I owe to Professor Frank my life-long interest in Einstein and his life and times, which have led to this book. I dedicate it to his memory.

1
The Prehistory

⌐ THE SCIENCE OF MECHANICS

Absolute, true, and mathematical time of itself, and by its own nature, flows uniformly on, without regard to anything external. It is also called *duration*.

    Relative, apparent and common time, is some sensible and external measure of absolute time (duration), estimated by the motions of bodies, whether accurate or in equable, and is commonly employed in place of true time; as an hour, a day, a month, a year....
—Isaac Newton

Our study of the prehistory of relativity begins with Galileo Galilei who was born in Pisa in 1564. We shall focus on one paragraph in one of his books, *Dialogue Concerning the Two Chief World Systems*. When he published it in 1632, he must have known that there would be trouble. He had brought the manuscript to Rome two years earlier to get permission from the Church to publish it. When this was not rapidly forthcoming, he returned to Florence, where he was living, and published it anyway with a Florentine *imprimatur*. Not only that, but he had written it in vernacular Italian as opposed to Latin, so that it could be widely read. The "world systems" in question are the Ptolemaic and the Copernican.[1] Ptolemy—Claudius Ptolemaeus—was an Alexandrine, probably of Greek origin, who lived in the second century BC. His astronomical system was a response to two apparently discordant requirements. On the one hand, he had inherited the notion from Aristotle that the heavenly objects, being made out of a different "essence" than earth, air, fire, and water, must move around the Earth in uniform circular motions while attached to crystalline spheres. The second requirement was that this system describe what one actually observed. This came to be called "saving the appearances." One of the "appearances," when it came to planetary motion, was that periodically, as seen from the Earth, planets go backward in their orbits—something that is known as "retrograde motion." To deal with this, Ptolemy introduced a remarkably ingenious system, which he adumbrated in his book *Almagest*. To take the simplest case, imagine a "virtual" planet that moves in a uniform circular motion around the Earth. Around this virtual planet the actual planet moves with a uniform circular motion. The combined orbits will show periodic retrograde motion. You can try this out by making the circles. In fact, by adding up circles you can simulate any observed planetary motion if you are willing to add up enough of them.[2]

---

[1] Galileo courted additional trouble by ignoring a third system that was currently in favor by the Church. In this system, invented by the Danish Astronomer Tycho Brahe, the Earth was at rest with the Sun in orbit around it, while the planets were in orbit around the Sun.

[2] Mathematically speaking, what Ptolemy did, unknown to him, was to generate what we would call a kind of Fourier analysis of the motion — an expansion in

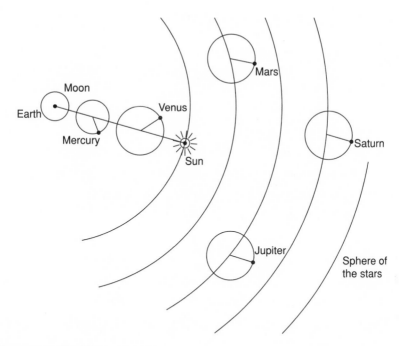

Figure 1.1. The Ptolemaic system.

Ptolemy used some fifteen for the Moon and planets.[3] Figure 1.1 is a rough idea of how it worked.

The second world system in the dialogues is the Copernican. Copernicus had presented this in his great book *De revolutionibus orbium caelestium*, which was published in 1543, the year of his death. People who have not actually studied what Copernicus wrote often misunderstand what he was proposing. What is usually recalled is that Copernicus moved the Sun to the center of the planetary system and made it stationary, with the Earth in motion. But, he also employed uniform circular motions and needed epicycles–even **more** than Ptolemy; in fact, some eighteen (see Figure 1.2). Figure 1.2 shows an additional complexity of the scheme, namely a displacement of the centers of the circles.

---

trigonometric functions. If you keep enough terms in the Fourier series, you can reproduce the original motion to any accuracy.

[3]I would like to thank Owen Gingerich for helpful communications.

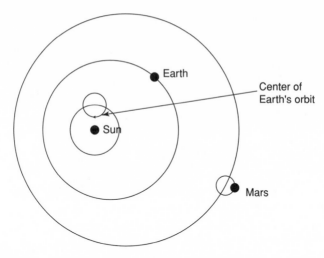

Figure 1.2. The treatment of Mars on the Copernican system. Thanks to Owen Gingerich for the drawing.

The solar system that is often depicted as "Copernican," with its elliptical orbits, as opposed to the uniformly moving crystalline spheres, was actually the discovery of Galileo's contemporary Johannes Kepler, whose diagram of the Martian orbit—an ellipse inside a circle for comparison, is shown in Figure 1.3. About the only thing it has in common with Copernicus is the resting Sun.

Galileo's concern in the "dialogues" was to show that the motion of the Earth, which is at the heart of the Copernican or Keplerian system, does not lead to absurdities. In the book, the dialogues, which take place over four days, are among three people. The setting, Galileo tells us, is in the palace of one Sagredo, who was modeled after a personal friend. Sagredo acts as the host and intelligent layman. Then there is Salviati, also modeled on a real person. Salviati, who is Galileo's stand-in, takes the Copernican side of the debate. Finally, there is Simplicius, an Aristotelean *pure et dure*, who, as one might imagine from the name, gets the worst of all the arguments. By this time the Aristotelean world view had become Church doctrine. So no matter how much he denied it, Galileo was challenging the Church. Indeed, in 1633, not long after the dialogues were published, he was summoned to Rome to face the

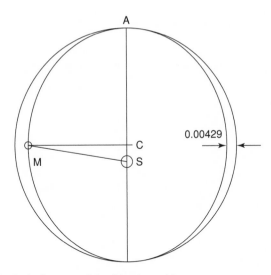

Figure 1.3. Kepler's diagram of the Martian orbit.

Inquisition. He returned to Florence a broken man and died there in 1642, the year Newton was born. The purpose of the dialogues, as I read them, is not to present the details of the Keplerian solar system. Indeed, Galileo's only interest in Kepler seems to have been to request from him additional proofs of the Earth's motion. In 1610, Kepler received from Galileo a copy of his book *Siderius nuncius*, which described his telescopic discoveries, such as mountains and craters on the moon and a system of moons revolving around Jupiter, all of which showed that the heavenly bodies were not so different from the Earth. Kepler was able to confirm these observations with a borrowed telescope. The purpose of the dialogues is rather to show that the objections that were being made to a moving Earth, at least the scientific objections, did not stand up to scrutiny. It is in this context that they begin our preamble to Einstein.

On the second day, Sagredo makes the following observation, "Ptolemy and his followers produce another experiment like that of projectiles, and it pertains to things, which separated from the earth, remain in the air a long time, such as clouds and birds in flight. [For these purposes projectiles also fall into this category.] Since of these it cannot be said that they are carried by the earth, as they do not adhere to it, it does not seem possible that they could keep up with its swiftness, rather it ought to look to us as they were being moved very rapidly westward." Why, in short, were objects aloft in the air not left behind by the moving Earth? This very reasonable concern provokes an extensive response from Salviati. In the course of it Salviati—Galileo—presents the following simple but extraordinarily profound insight. Einstein liked to use trains in his examples. Galileo used a sailing ship. Here is what he writes,

> Shut yourself up with some friend in the main cabin below deck on some large ship, and have with you some flies, butterflies, and other small animals. Have a large bowl of water with some fish in it; hand up the bottle that empties drop by drop into a narrow-mouthed vessel beneath it. With the ship standing still, observe carefully how the little animals fly with equal speed to all sides of the cabin. The fish swim indifferently in all directions; the drops fall into the vessel; and in throwing something to your friend, you need throw it no more strongly in one direction than another, the distances being equal; jumping with your feet together, you pass equal spaces in every direction.

Now comes the crucial observation.

> When you have observed all these things carefully (though there is no doubt that when a ship is standing still everything must happen this way), have the ship proceed with any speed you like so long as the motion is uniform, and not fluctuating this way and that. You will discover not the least change in all the effects named, nor could you tell from any of them whether the ship was moving or standing still.

This is the first time in which what we call a "relativity" principle was described explicitly. We can restate Galileo's charming folkloric presentation somewhat more austerely as follows: In no experiment done in a uniformly moving system, does the speed of that system with respect to any other uniformly moving system, play a role. In other words, for purposes of any experiment, we can take our uniformly moving system to be at rest. It is called a "relativity principle" because, as far as uniform

motions are concerned, all that is measurable is the "relative" velocity
of one system "relative" to another. In a uniformly moving train or car
or plane, we only know we are in motion when we view the tracks, the
road, or the ground. If we want to be perverse about it, we can say that
we are at rest and these reference systems are the one's in motion. This
seems totally innocuous and commonsensical–but wait until we come to
Einstein. In the meanwhile let us see how the principle is realized in the
mechanics of Newton.

Newton's mechanics were laid out formally in his seminal book
Phliosophiae naturalis principia mathematica, which was first published
in 1687. Newton not only created new science, but a new scientific
paradigm. He invented what we think of as theoretical physics. You
start with some general principles that aid you in formulating a set of
equations. You solve these equations as best you can and check the result against experiment. No one prior to Newton had done things in this
way. For example, Kepler did not try to *derive* the elliptical planetary
orbits. He showed empirically that this was how the planets move. In
the *Principia*, Newton was able to derive the planetary motions from a
few general principles. I will now present some of them, beginning with
Newton's "second law," not quite as stated in the *Principia*–we will come
to that shortly–but in a form that will be familiar to many of you. I will
write the formula and then say what the letters mean, or at least crudely
what they mean. A little later I am going to critically analyze these equations in the sprit of Einstein's influential contemporary Ernst Mach,
the Austrian physicist-philosopher whose book *The Science of Mechanics*
played a very important role in Einstein's thinking.

Put in the simplest language, Newton's Second Law says that the
acceleration an object experiences is proportional to the force applied
to it. The constant of proportionality is the mass of the object. In short,
$F = ma$. I am assuming for the moment that we have some general idea
of what these terms mean. When I come to Mach's critique, it is this we
will have to examine. I want to focus on the acceleration. To say that
an object is accelerated is to say that its motion has been changed. This
can mean that its direction has changed or that while moving in a given

direction it changes its speed or both. To simplify, I will suppose that the motion is along a one-dimensional straight line. This simplification will not change anything essential. Let us consider two times–"initial" and "final"–which we denote by t(initial) and t(final). If the motion is accelerated, the speeds at these times, v(initial) and v(final), are different. We can form the quantity (v(final) − v(initial))/(t(final) − t(initial)). This gives a measure of how much the speed has changed in the interval in question. What Newton did was to allow the interval to get smaller and smaller so that, in the limit of an infinitesimally small interval, we have the ratio at a single time somewhere in the middle. This limit is how the acceleration at some arbitrary time is defined. This limiting process is an example of the differential calculus which Newton invented for this purpose. Now I want to persuade you that Newton's law as we have stated it obeys the relativity principle, at least in this example. The argument can be generalized to any motion.

First, consider the right-hand side of the equation. How would this look to an observer in uniform motion with a speed that we shall call v(relative)? The speed, v(relative), in our one-dimensional example can be positive or negative. Now, common sense tells us that to rewrite the equation for the acceleration from the point of view of the moving observer, we should simply add v(relative) to whatever velocities we have at hand. Thus, the numerator in the Newton's law equation becomes v(final) + v(relative) − v(initial) − v(relative). We see that v(relative) has canceled out so that the numerator takes exactly the same form in both systems. Common sense also tells us that the times do not depend on v(relative) so that the denominator does not change either. In fact, both of these common sense observations turn out to be wrong, as we will learn in the next chapter when we discuss Einstein's relativity. What about the left-hand side of the equation? The forces that Newton considered–primarily gravitation–do not depend on the velocities. For example, the gravitational attraction between two objects decreases as the square of the distance—as $1/d^2$–which does not depend on which uniformly moving systems you view these objects from. Later in this chapter we will consider electromagnetic forces that do depend on velocities. This introduces a new element into the discussion of relativity. It is not an accident that Einstein called his relativity paper "On the Electrodynamics of Moving Bodies." But we see here how relativity is built into Newtonian mechanics. In our example, the equation has exactly the same form in a system at rest and in a system moving uniformly with respect to it. This is something that was after Einstein's relativity called the "covariance" of the equations—the fact that they take the same form in different reference systems.

Newton's mechanics were so successful that for the next two centuries the foundation on which they were based was not critically examined. The first person to do this, at least the first person to do it whose work had an impact, was the above-mentioned Mach. Before I explain Mach's objections let me say a bit about him. He was born Ernst Walfried Joseph Wenzel Mach on 18 February 1838, in the Austro-Hungarian town of Chirlitz. This makes him a good deal older than Einstein who, remember, was born in 1879. Nonetheless, the two men met in 1912, in Vienna, a meeting that Professor Frank arranged and attended. Einstein and Mach, Professor Frank recalled, discussed the "existence" of atoms, to which we shall devote the third chapter in which I shall discuss why Mach thought that atoms did not exist. At the age of nine, Mach was enrolled in a Benedictine Gymnasium near Vienna. The fathers there rated him as "*sehr talentlos*"—more or less hopeless. He was then tutored by his own father who used to shout at him imprecations like "Norse brains" or "Head of Greenlander." Mach decided that he would become a cabinetmaker and move to America and, indeed, for two years he was apprenticed to a cabinetmaker. If you read Mach's great polemic book, *The Science of Mechanics*, you will be struck by the illustrations of mechanical devices that look like they could have been built by a cabinetmaker.

At the age of fifteen he returned to the Gymnasium and later wrote,

With respect to social relations and the like I must have seemed extremely immature and childish. Apart from my slight talent in this direction, this is to be explained to some extent by the fact that

> I was fifteen years old before I ever engaged in social intercourse, particularly with students of my own age.... At the beginning things did not go especially well, since I lacked all of the school cleverness and slyness which first have to be acquired in these matters.

Despite this rather unpromising start, Mach was able to enter the University of Vienna in 1855, where he received his PhD five years later working on what seems to have been experimental aspects of electricity. Mach, by his own admission, never had a strong background in mathematics and never did any significant theoretical physics. After taking his degree, Mach became a *Privatdozent* at the university, which allowed him to lecture. The students paid him directly. Professor Frank held the same position a few decades later. Mach really earned his living by giving popular and semipopular lectures–especially to medical students–that were later published. After a period at the university in Graz, in 1866, Mach became a professor at the German University in Prague. Both Einstein and Frank became professors there. One of Mach's early interests was the Doppler shift. We are familiar with it because we hear the shift in pitch of approaching sirens and train whistles–fairly rapidly moving vehicles. But, when the Austrian physicist Christian Doppler proposed it in 1842, on theoretical grounds, there was a great deal of skepticism that lasted for many years. One of Mach's own professors, Joseph Petzval, claimed that it was impossible because of something he called the "law of conservation of the period of oscillation." In 1860, Mach built a simple apparatus to demonstrate it. It consisted of a long tube that was free to rotate around a central axis. A sound was produced in the tube by forcing wind through it. If one was stationed in the plane of rotation of the tube one heard the shift, while if one stationed oneself on the axis of rotation, it disappeared. (As we shall see, this is a feature of the "classical" Doppler shift which no longer holds in Einstein's theory.) Even so, in 1878, Mach had to persuade a group of teachers and students to sit on a hill overlooking some railroad tracks and listen to whistles of approaching trains. Afterwards, they signed a document as a testament to what they had heard.

In presenting his critique of Newton's formulation of his mechanics, Mach did not argue that the physics was actually wrong and needed to be replaced. His point was that Newton had assumed implicitly, or explicitly, metaphysical or theological doctrines that made his theory unscientific. What Mach wanted to do was to rid the theory of this baggage—"harmful vermin"—as Einstein later referred to it. The *Principia* begins with a series of definitions followed by a set of Laws. The first definition already shows the problem. It reads:

> *Definition* I. The quantity of any matter is the measure of it by its density and volume conjointly. The quantity is what I shall understand by the term *mass* or *body* in the discussions to follow. It is ascertainable from the weight of the body in question. For I have found, by pendulum experiments of high precision, that the mass of a body is proportional to its weight; as will hereafter be shown.

Two things are evidently wrong with this definition. Density is mass per unit volume, so the definition is circular. Secondly, he confounds mass with weight. Didn't Newton realize that, for example, on the Moon you would weigh about a sixth of your earthly weight, even though your mass would not have changed? However, one can see, that for many practical purposes, this definition might serve, even though it is fundamentally flawed. Mach writes,

> ... we do not find the expression "quantity of matter" adapted to explain and elucidate the concept of mass, since that expression itself is not possessed of the requisite clearness. And this is so, though we go back, as many authors have done, to an enumeration of the hypothetical atoms. We only complicate, in so doing, indefensible conceptions. ...

Mach is saying that it is no good trying to explain masses in terms of atoms, because these are themselves "indefensible conceptions. However, it is in terms of the ill-defined notion of mass, that Newton makes his second definition;

Definition II. Quantity of motion is the measure of it by the velocity and quantity of matter conjointly.

"Quantity of motion" is what we would now call "momentum." It is usually designated by the letter "p." What Newton is saying is that by definition the momentum p is equal to mv, $p = mv$, where "v" is the velocity and "m" is the mass. In terms of these definitions Newton then stated the Second Law.

Law II. Change of motion [i.e., of momentum] is proportional to the moving force impressed, and takes place in the direction of the straight line in which such force is impressed.

This statement of the Second Law is more general than the one I used before which was simply $F = ma$. Here, it is $F = \frac{\Delta p}{\Delta t}$, the change of momentum as a function of time. So long as the mass does not depend on time the two definitions are the same. If not, we must also take account of the time-dependent mass change. In Einstein's theory of relativity, as we shall see, the mass can depend on time, so we must use the Newtonian form of the Second Law. We must also use it, for example, in describing rocket propulsion as the mass of the rocket diminishes because the propellant is being expelled. Mach has no problem with the generalization. He does have a problem with the potential circularity of the law. There is no independent definition of "force." To see if a force has acted, and what its characteristics might be, we must have an independent definition of acceleration. From the way I have approached it earlier, it is clear that this comes down to having a completely reliable measurement of time. If we have a faulty clock—one that speeds up, or slows down, or even stops erratically, during the time interval within which we are trying to measure the acceleration—we can get all sorts of results. We might attribute these to an erratic clock, or to some new force. There would be no obvious way of telling the difference. The Second Law would lose its content.

Newton seems to have been aware of this. It is presumably why he felt that it was necessary to introduce "absolute" time, one of whose characteristics was precisely that it did not require measurement by clocks.

To this, Mach objects vigorously. He writes, "It would appear as though Newton in the remarks here cited [on "absolute" time] still stood under the influence of the medieval philosophy, as though he had grown unfaithful to his resolves to investigate actual facts"

He goes on to conclude,

> ... But we must not forget that all things in the world are connected to one another and depend on one another, and that we ourselves and all our thoughts are also a part of nature. It is utterly beyond our power to *measure* the changes of things by *time*. Quite the contrary, time is an abstraction at which we arrive by means of the changes of things; made because we are not restricted to any one *definite* measure, all being interconnected. A motion is termed uniform in which equal increments of space described by some motion with which we form comparison, as the rotation of the earth. A motion, with respect to another motion, be uniform. But the question whether a motion is *in itself* uniform, is senseless. With just as little justice, also, may we speak of an "absolute time"–*of a time independent of* change. This absolute time can be measured by comparison with no motion; it has neither a practical nor a scientific value; and no one is justified in saying that he knows aught about it. It is an idle metaphysical conception.

One can only imagine the sense of liberation that the young Einstein must have felt when he read these words.

Newton was much too great a scientist not to have understood that uniform motion had no absolute meaning, at least scientifically. He must have realized that the relativity principle was built into his theory. However, he felt that acceleration was something different. He was sure that absolute acceleration made sense. He gave two reasons for this; one scientific and the other theological. Mach's discussion of the former is one of the most noted portions of his book. Newton makes a distinction between "absolute" and "relative" motion. He writes,

> Absolute motion is the translation of a body from one absolute place to another absolute place; and relative motion the translation from one relative place to another relative place.... And thus we use in

common affairs, instead of *absolute* places and motions, *relative* ones; and that without any inconvenience. But in physical disquisitions we should abstract from the senses. For it may be that there is no body really at rest, to which the places and motions of others can be referred.

In short, Newton is claiming that these absolute motions should be meaningful in empty space. To support this notion he brings in the matter of circular motion whose effects, he claims, are manifestly absolute. He writes,

The effects by which absolute and relative motions are distinguished from one another, are centrifugal forces, or those forces in circular motion which produce a tendency of recession from the axis. For in a circular motion which is purely relative no such forces exist; but in a true and absolute circular motion they do exist; and are greater or less according to the quantity of the [absolute] motion.

Newton then clarifies this rather obscure pronouncement by citing an experiment. He writes,

For instance. If a bucket, suspended by a long cord, is so often turned about that finally the cord is strongly twisted, then is filled with water, and held at rest together with the water; and afterwards by the action of second force, it is suddenly set whirling about the contrary way, and continues, while the cord is untwisting itself, for some time in this motion; the surface of the water will at first be level, just as it was before the vessel began to move; but, subsequently, the vessel, by gradually communicating its motion to the water, will make it begin sensibly to rotate, and the water will recede little by little from the middle and rise up the sides of the vessel, its surface assuming a concave form. (This experiment I have made myself.)

In short, at first sight, it looks as if Newton has constructed an experiment in which the effects of rotation seem not to depend on any reference system. You could imagine performing it in empty space. Mach's response is to point out that Newton's experiment is not done in empty space. He writes,

> Newton's experiment with the rotating vessel of water simply informs us, that the relative rotation of the water with respect to the sides of the vessel produces *no* centrifugal forces, but that such forces *are* produced by its relative rotation with respect to the mass of the earth and the other celestial bodies. No one is competent to say how the experiment would turn out if the sides of the vessel increased in thickness and mass till they were ultimately several leagues thick. The one experiment only lies before us, and our business is, to bring it into accord with the other facts known to us, and not with arbitrary fictions of our imagination.

In 1916, Einstein published his theory of gravitation. In 1918, the Austrian physicist Hans Thirring, using the theory, considered what would happen if you had the equivalent of Newton's bucket at the center of a hollow sphere with a massive shell and you rotated the shell. He found that you would have the same affect at the center as if you had rotated the bucket. In short, Mach was right. You cannot have any idea of what would happen in empty space, because space is not empty. Newton, as I have mentioned, was a deeply religious man. For him, the distinction between science and religion did not really exist. He noted, that while absolute motions might not be observable to us, they were well-defined in the *sensorium* of God. Newtonian mechanics rested on a theological and metaphysical base which, because of its vast success, had been overlooked.

Next we turn to light and electricity.

The Aether

> Upon considering the phenomena of the aberration of stars, I am disposed to believe that the luminiferous aether pervades the substance of all material bodies with little or no resistance as the wind passes through a grove of trees.
> —Thomas Young

The first person to introduce into physics what later came to be called the "aether" was the French polymath René Descartes. Descartes, who was born in France in 1596, lived much of his life in Holland and died in Sweden in 1650. His life overlapped with that of Newton, who used his mathematics and was derisory about much of his physics. For example, Newton used Descartes algebraic characterization of an ellipse–the one we use today. Descartes took Galileo's encounter with the Church as a warning and concocted a kind of mixed Copernican universe. The Earth was at rest at some kind of vortex which then went around the Sun. Descartes and Kepler had in common the idea that this orbital motion had to be explained by a "hands on" set of influences. In Kepler's case it was a magnetic force emanating from the Sun and for Descartes it was the aether. This medium, which was not directly detectable, had vortices and swirls (see Figure 1.4). They were forever changing and when the Earth got caught in a vortex it was swept along like a boat in a turbulent stream. This picture avoided what was thought to be the intolerable notion of "action at a distance"–an influence propagated, perhaps instantaneously, from a far off object, such as the Sun, to the Earth with nothing like a discernible push or pull. One of the manifestations of the greatness of Newton was that he resisted the temptation to "explain" gravitation. He wrote down the law and then derived the consequences.

Over the next century the aether, or aethers, became a "theory of everything." There was an electrical aether that transmitted electrical influences. There was a heat aether — caloric — that was the medium that transported heat. Indeed, it **was** heat. It was well understood that sound was transported in media like the air–a material aether. Beginning in the time of Newton, there was also a "luminiferous aether" that, in one form or other, transmitted light. Indeed, there were two theories of light. One of them argued that light was transmitted, like arrows through the air, by particles that penetrated the aether. The other held that light was in fact a series of vibrations of the aether, resembling sound, but moving very much faster. It was known that light, unlike sound, could be transmitted through whatever vacua could be made in the laboratory–but these

Figure 1.4. Descartes' aether had vortices and swirls.

"vacua" were also permeated by the aether. Newton's view of this is very instructive. He was not entirely sure what light was. He wrote that light is

> something of a different kind, propagated from lucid bodies. They, that will, may suppose it an aggregate of various peripatetic qualities. Others may suppose it multitudes of unimaginable small and swift corpuscles of various sizes, springing from shining bodies at great distances one after another; but yet without any sensible interval of time, and continually urged forward by a principle of motion, which in the beginning accelerates them, till the resistance of the aethereal medium equals the force of that principle, much after the manner that bodies let fall in the water are accelerated till the resistance of the water equals the force of gravity.... But they, like not this, may suppose light any other corporeal emanation, or any impulse or motion of any other medium or aetherial spirit diffused through the main body of aether, or what else they can imagine proper for this purpose. To avoid dispute, and make this hypothesis general, let every man here take his fancy; only whatever light be, I suppose it consists of rays differing from one another in contingent circumstances, as bigness, form, or vigour.

Newton, as I will now explain, had good scientific reasons for rejecting both the particle and wave theories of light as they were then being expounded.

The particle theory appeared to make a clear prediction of how objects cast shadows. They should be sharp since the object presumably just blocked the particles. However, an Italian physicist named Francesco Maria Grimaldi, whom Newton referred to as "Grimaldo," had made the discovery that these objects cast shadows that were too large—larger than they "ought" to be—as Newton put it. He verified this phenomenon—"diffraction"—for himself. This seemed very difficult to reconcile with a particle theory of light. To understand Newton's objections to the wave theory we have to understand what was on offer, something very different to the light waves we are familiar with. There was an understanding of how sound propagates in a medium—some of this due to

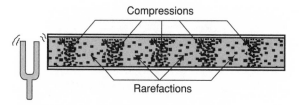

Figure 1.5. A tuning fork emitting sound waves in, say, air.

Newton himself. I will put it baldly and then explain. Sound propagates as a longitudinal pressure wave. What does that mean?

See what happens in Figure 1.5. As the tuning fork vibrates, the vibrations push the adjacent air molecules. As shown, these move to the right. Others can move in other directions, but for simplicity let us consider the right-moving ones alone. Because of the motion of the air and the tuning fork there will be regions of compression and rarefactions. These regions will assume a periodic character reflecting the periodic motions of the tuning fork. Newton described this picturesquely as "fits of easy reflection" and "fits of easy transmission." If you plot the pressure measured at any point as a function of time it oscillates up and down periodically. But the periodicity of the wave itself consists of the pattern of compression and rarefaction that reflect the oscillations of the tuning fork. Later I will discuss light waves more carefully, but here I will just note that they oscillate in a plane at right angles to their movement, while sound waves oscillate in the direction of their movement, hence the term "longitudinal." The way a wave-length is assigned to these longitudinal waves is to measure the distance from, say, one compression to the next. The frequency of such a wave is the number of compressions that pass a given point per second. The product of the frequency and the wave length measures the speed of propagation. I will shortly explain how using waves explains optical phenomena like diffraction—something that Newton seemed to understand. But here I want to indicate the problem that Newton had with the wave theory of light. It came from an unlikely source—Iceland.

Sometime in the 17th century a sailor brought back from Iceland to Copenhagen, beautiful crystals—really cleavage fragments of

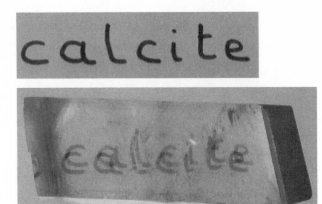

Figure 1.6. An example of the optical effect when viewing an object through "Iceland Spar" crystal. (Photograph courtesy Joanna Edkins)

calcite–which were called "Iceland Spar." He had collected them in the Bay of Röerford.

A Dane named Erasmus Bartholin observed a remarkable optical effect. Small objects viewed through such a calcite crystal appear doubled. Each of the rays seemed to have different optical properties (Figure 1.6). The first person to make a significant study of this was the Dutch physicist Christiaan Huygens. He was born in 1629 and died in 1695, so he also overlapped with Newton. Huygens was a brilliant scientist. He was the first person to make a quantitative estimate of the distance to a star–Sirius. He assumed that it had the same intrinsic luminosity as the Sun and found it to be 2.5 trillion miles away, as opposed to the actual distance which is twenty times farther, since Sirius is more luminous than the Sun. His most important work was in the support of the wave theory. He had a variety of objections to the particle theory including the fact that light rays can cross over each other without evident hindrance. Indeed, the most characteristic property of waves is how they interact. When two waves meet they interfere to produce a resultant wave. If crests meet crests then the waves are enhanced. If crests meet troughs the waves can annihilate each other. Particles, at least classical particles, do not annihilate each other. Huygens invented a method for calculating the future progress of a light wave from its

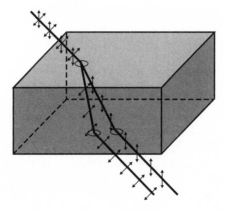

Figure 1.7. The two refracted rays passing through the Iceland Spar crystal are polarized with perpendicular orientations.

characteristics at any prior time. I will come to it in a moment, but first I want to discuss his observations with the Iceland Spar.

Huygens had the ingenious idea of using two of these crystals in series. The first crystal split the beam into two. But Huygens noticed that the second crystal did a variety of things if he let a single beam from the first crystal go though it. Sometimes the single beam would be split in two, and sometimes it would not. Sometimes, depending on the orientation of the two crystals–how they had been rotated with respect to each other–the second crystal would not transmit the beam at all. Figure 1.7 shows an anachronistic diagram of what happened in the first crystal. It is anachronistic because we know the answer, which Huygens did not. It is what we call "polarization." The little arrows show how the light ray oscillates compared to the direction in which it propagates. What happens is that the first crystal acts as a polarizer and the second crystal as an analyzer that only allows transmission of light that has the proper orientation of its polarization. Huygens did not know this. It was only fully understood in the nineteenth century.

When Newton heard about Huygens's result, he correctly drew the conclusion that light could not be a longitudinal pressure wave like sound. He does not seem to have considered any other kind of wave as a possibility.

Huygens wavelets

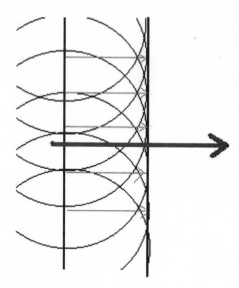

Figure 1.8. Huygens wave construction.

The Huygens wave construction is the proposition that each point on a wave front, at a given time, acts as a source for the future wave. Suppose, as in Figure 1.8, that the wave fronts are planes — something we call a "plane wave." At each point on one of the planes we draw a circle, part of which sticks out to where the wave will go. If we fix the radii of the circles we fix the next moment in time. We then draw the envelope of all these circles where they stick out into the future–a figure tangent to each of the circles. In this case it will be a plane, as Figure 1.8 shows.

Huygens did not have the mathematics to work out his construction in complex cases–this was left to the early 19th-century mathematical physicist Augustin Fresnel–but he did see how to use it to explain many of the optical phenomena he had observed.

Nonetheless, there were, even in the early 19th century, partisans of both the particle and the wave theories of light. The matter was settled to everyone's satisfaction by the work of the British polymath Thomas Young. We will meet Young again on the third chapter in which we discuss the "existence" of atoms. He was the first person to make a quantitative measurement of the sizes of molecules. We would also meet him if this was a book on Egyptian hieroglyphics. Using material brought back from Egypt by the French, Young was able to make the start of a Greek-demotic dictionary—demotic being an early Egyptian written language. From this he was able to decipher a few of the hieroglyphs which had demotic translations. The job, using Young's work, was completed by the French linguist Jean François Champollion. Young, who could read at age two—he was born in 1773—and had read the Bible twice by age six, had already started the study of Latin. By the time he was twenty, he had mastered at least a dozen languages. Young began his career as a doctor but soon wandered into science. At the turn of the century he was defending the wave theory of light. By 1801, he had done the first of his experiments that seemed to prove it once and for all (see Figure 1.9).

You notice that a plane wave is incident on a barrier with two slits through which the light can pass. If the light consisted of particles they would pass through the slits, when they do so, and leave simple spots on the detector. But this is not what Young observed. He found a series of fringes. He showed how this could be understood by the interference of waves. Figure 1.9 shows two waves interfering which will produce bright

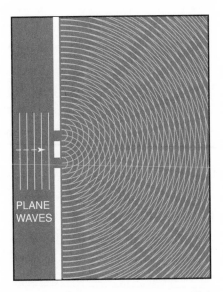

Figure 1.9. Young's demonstration of the wave-nature of light.

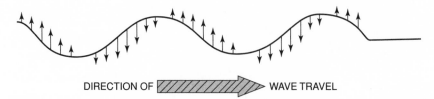

Figure 1.10. Light is entirely transverse, as shown by the direction in which vibrations travel.

lines on a detector. In 1810, Young learned of an experiment in which reflected light was polarized similar to what happens in the Icelandic Spar. Young made a suggestion that went part of the way towards an explanation. He said that a light-ray might have a partially transverse part. Later Fresnel showed that light was entirely transverse, meaning that the direction of vibration was always at right angles to the direction of propagation (Figure 1.10).

This posed a profound problem for the aether theorists with which they were occupied for the rest of the century.

The reason that sound is a longitudinal wave is because the media it travels through have little or no elastic resistance. It is like shooting bullets through fog. There is no resistance to deformation. But light waves are transverse which means that the aether must be some sort of elastic **solid**! This seems quite mad. We are, it was pointed out, moving through this hypothetical aether all the time, to say nothing of the other planets, and we do not sense any resistance, and the planetary orbits are not affected by it. It turned out that not much was known about elastic solids. This dilemma inspired a mathematical development whose fruits we enjoy to the present day, even though the motivation has long been lost. For example, a British theoretical physicist named George Stokes presented a theory that had analogies to the behavior of substances like shoemaker's wax, which were plastic but allowed bodies to pass through them. Perhaps the aether was an extreme form of this which had these special properties because of the great speed of light and the small wavelengths. Visible light has wave lengths typically of about fifty thousandths of an inch, while the shortest wavelength of sound audible to us is about a half inch.

As we now shall see, by the end of the century, experiments had been performed on light which had sufficient accuracy so that they should have been able to detect the speed of the Earth through the aether. These experiments detected nothing. This provoked a major crisis in physics which was unresolved as the century closed.

↰ CONTRACTIONS

Einstein, following a suggestion originally made by Poincaré, then proposed that *all the physical laws* should be of such a kind that they *remain unchanged under a Lorentz transformation.*
— Richard Feynman

In the early 1970s, I wrote a brief biography of Einstein. He had died on April 18, 1955, but his house in Princeton at 112 Mercer Street was still occupied by his stepdaughter Margot, and by Helen Dukas, who had been his secretary since the 1920s and had come here from Germany in 1933 with Einstein. Professor Frank had known Miss Dukas for years, so when I went to work at the Institute for Advanced Study at Princeton in 1957, it was easy for me to meet her. When I began doing research on my book, I asked if I could visit the house, and Miss Dukas was kind enough to agree. Einstein lived in a small apartment on the second floor, which was pretty much as he had left it. Miss Dukas said there were a few more plants. The chair and table on which he worked were still there, as was his small library of general books, which were pretty serious. I noticed Frazer's *Golden Bough* and Gandhi's autobiography. On the wall I noticed something that looked a little odd. There was an etching of the Scottish physicist James Clerk Maxwell, whose work we are about to discuss. Next to it was a frame of about the same size with a bit of modern art. I asked Miss Dukas whether Einstein had chosen the art. She said that what had happened is that there had been an etching of Newton which had come out of its frame. This happened after Einstein's death and it had been replaced by the modern art. I found this very amusing. As we shall see, Einstein's theory of relativity leaves Maxwell intact, but replaces Newton.

The 17th century is often referred to as the age of Newton. I think the 19th century might be called the age of Maxwell. It is not that the two men did not have important predecessors and contemporaries, but they towered over everyone else. In Newton's case there were Galileo and Kepler, and the mathematics of Descartes, to mention a few. In Maxwell's case, these people are probably best known to most of us because physical units that we still use are named after them. Here are a few examples. In the late 18th century, the French physicist Charles-Augustin Coulomb discovered that forces between charges fall off as the inverse square of the distance between them–like the force of gravitation. In 1819, the Danish physicist Hans-Christian Ørsted found that electric currents produce magnetic actions such as moving compass needles. This was the first hint that electricity and magnetism might be related phenomena. About the same time, the French physicist André Marie Ampère found the law of magnetic force between two electric currents. And then there is Michael Faraday. Faraday, was a poor boy, born near London in 1791, who worked as a bookbinder's apprentice. His introduction to science was in reading some of the pages he was binding. In 1812, one of the customers gave Faraday tickets to hear lectures by the chemist Humphry Davy. Faraday took detailed notes and sent Davy a bound copy. This eventually led to a position as Davy's assistant. Davy took him on a trip to Europe where they met the scientists who were studying electricity and magnetism. When Faraday returned to London, he repeated all the experiments he had seen, and then began his own. These ultimately led to Faraday's idea of lines of force. Using iron filings, he mapped out how magnetic forces varied in space. This, in turn, led to the notion of fields of force which became the basis of Maxwell's work. Faraday never learned much mathematics, but the one theory he said he understood was Maxwell's. Maxwell gave mathematical expression to Faraday's discoveries. It is the mathematics we still teach and use.

Maxwell was born in Edinburgh in June of 1831. He came from a prominent Scottish family and ultimately inherited 1500 acres of farmland in southwestern Scotland. He did much of his scientific writing on his estate. This began at the age of fourteen, when he published a paper on a method for drawing a perfect oval. His next papers were on geometrical subjects. Then he turned to optics. He studied at Cambridge and began working on electricity shortly after he graduated in 1854. His lifetime work on this culminated in his great book, *Treatise on Electricity and Magnetism*, which was published in two volumes in 1873. It was to this subject what Newton's *Principia* was to the science of mechanics. Volumes have been written about Maxwell's work in physics, which covered the entire range from a theory of Saturn's rings to seminal work in statistical mechanics. We shall discuss a bit of the latter in our chapter on atoms. Here, I want to focus on a small part of his electromagnetic work.

Maxwell first derived his equations using a baroque model of the aether in which, for example, rotating vortices represented magnetic fields. How he was able to see through the details of this model to the underlying equations is a mystery to me. In any event, by the time he came to write his book, the model had disappeared. The equations remain.

When contemplating his model, Maxwell asked himself how would a disturbance in this aether propagate. How would the electric and magnetic fields generated by this disturbance evolve in time. He discovered that they would propagate as transverse waves. In his model there was an expression for the speed of the wave in terms of measurable electromagnetic parameters. When he put in the experimental values for them he discovered that the speed that emerged was the speed of light, or very nearly. His number was 3.1×10^{10} centimeters per second, while the best measured number was, at the time, 3.15×10^{10} centimeters per second. In the next chapter I will describe these experiments. Maxwell drew the tentative, and then very radical, conclusion that light was an electromagnetic wave. He did not suggest a way of testing this. This was left to the German physicist, Heinrich Rudolf Hertz. Hertz published his experiment in 1887, by which time Maxwell had been dead for several years. He died of cancer in 1879, at the age of forty-eight. What Hertz did was to produce electromagnetic waves using oscillating electric currents. One circuit was used as an emitter and another as a detector. In one instance he put the detector at the other end of a large room and watched as sparks were created when the wave from the emitter arrived. By moving the detector around he could measure the wavelength and, ultimately, the velocity of the waves–again finding the speed of light. That light was an electromagnetic wave was now demonstrated.

Not long before his death, Maxwell wrote a letter to the American Astronomer David Peck Todd, who was the director of the Nautical Almanac Office in Washington, DC, which, despite its name, did some basic work in astronomy. Maxwell had wondered if the speed of the Earth through the aether was measurable. He had in mind using light coming from the satellites of Jupiter. He explained the reason why he thought using terrestrial light sources would not work. His argument is a template for our future discussion which will get us closer to Einstein's relativity. So we should pay special attention to it. In the aether theory, the speed of light is determined by the physical properties of the aether, just as the speed of sound is determined by the properties of the medium in which the sound is propagated. Once you introduce light in the resting aether, it propagates with a speed that I shall call–using the common convention– "c." By the way, Maxwell used "v" for the speed of light and this was adopted by Einstein in his 1905 paper. A few years later he switched to "c." This is the speed that would be measured by any observer at rest in the aether. But, suppose you are not at rest. Common sense tells us that if we are moving towards the advancing light with a speed v the light would appear to be moving with respect to us with a speed $c + v$. On the other hand, if we are moving away from the light the speed would be $c - v$. This suggests a way of measuring v. We could set up a light emitter somewhere and then go a distance "L" and time how long it takes for the light to arrive there. This time depends on our motion with respect to the aether. If we are moving towards the light the time is $\frac{L}{c + v}$, while if we are moving away from the light the time is $\frac{L}{c - v}$. You may well object that, a priori, we do not know in which direction we are moving in the aether. I agree, but to get around this let us time the flight now, and then six months from now, when the Earth is moving in the opposite direction. We should see a difference if the aether theory is right. There is nothing wrong with this method in principle. In practice, it won't work, at least for the clocks that were available in the 19th century.[4]

To see why, recall that the orbital speed of the Earth is about 3×10^6 centimeters per second. What measures the size of the effect above for a one-way trip, is the ratio of this speed to that of light $-v/c-$ which is about one ten thousandth. Putting the matter algebraically, $\frac{L}{c-v} = \frac{L}{c}\left[\frac{1}{1-\frac{v}{c}}\right] \approx \frac{L}{c}\left[1+\frac{v}{c}\right]$.[5] We see that the correction to the time it would take if we were at rest in the aether, L/c, is of order v/c. The speed of light is so fast that, with the clocks that were then at hand, you would never observe this effect. For all practical purposes the transmission of light between two points within a moderate distance from each other is instantaneous. It may have occurred to the reader—it certainly occurred to Maxwell—that there is a simple way to double the path-length and perhaps make the passage of the light a little less instantaneous. You simply put a mirror at the distance L, and time the round trip from the source and back again. With our formulae we can see that the total time for this round trip is

$$\frac{L}{c-v} + \frac{L}{c+v} = 2cL\frac{1}{c^2-v^2} = \frac{2L}{c}\frac{1}{1-\frac{v^2}{c^2}} \approx \frac{2L}{c} \times (1+\frac{v^2}{c^2})$$

Thus, this effect is of order v^2/c^2 relative to the time if we were at rest which is $2L/c$. We recall that v/c is one ten-thousandth, which means that v^2/c^2 is one hundred millionth! Maxwell argued, that since any practical terrestrial measurement would have to involve a round trip for the light, such a measurement was out of reach. As it happened, that while Maxwell had the correct address on the envelope of his letter, he had addressed it to the wrong person. He should have addressed it to a young naval officer named Albert Abraham Michelson, who was then spending a good deal of time at the Nautical Almanac Office.

---

[4]There are subtle issues involved in such time of flight measurements that I am ignoring. Implicitly they assume that distant clocks can be synchronized. This is just the sort of issue that Einstein raised in his discussion of the relativity of time, the subject of the next chapter.

[5]In general for x much smaller than 1 we have $\frac{1}{1\pm x} = 1 \mp x$.

In 1907, Michelson became the first American to win the Nobel Prize in any science–physics in his case. Like many subsequent American winners of the Prize, he was not born in this country. He was born on December 1852, in what is now Strzelno, Poland. When he was three, fleeing from anti-Semitic pogroms, he left Poland with his mother and father for the United States. After arriving in New York, the family set sail for California, where they heard that gold had been discovered. Since there was no Panama Canal, this involved traversing the Isthmus of Panama by mule and canoe. They ended up at Murphy's Camp, a mining settlement, at the foothills of the Sierra Nevada. Michelson's father set up a store to sell things like picks and shovels to the miners. Michelson received a rudimentary education in a local school and then was sent to San Francisco for his high school education. While there, he began studying optics on his own. His father found a newspaper item which said that there was a possible opening at the Naval Academy in Annapolis. Michelson applied, and failed to get the appointment. However, he had heard that the President–Grant in this instance–had ten appointments-at-large at his disposal. Michelson made the cross-county trip to Washington alone and, remarkably, had an interview with Grant, only to be told that the appointments had been made. But he was encouraged to go to Annapolis in case one of the appointees did not qualify, which is what happened.

Michelson proved to be a brilliant student in science, less so in seamanship. After the obligatory sea duty, Michelson returned to Annapolis where he was asked to assist in a physics course. While preparing, he invented a method of improving the existing measurements of the speed of light. He found the value 299,910 kilometers per second, which was some two hundred times more accurate than any previous measurement. This caught the attention of Simon Newcomb, a very distinguished astronomer, who was then the director of the Nautical Almanac Office. Newcomb invited Michelson to join him in an even better measurement of the speed of light. It was during this period when Maxwell's letter to Todd arrived. Michelson, who was shown the letter, decided that he would try to devise a method to measure the aether speed to a few

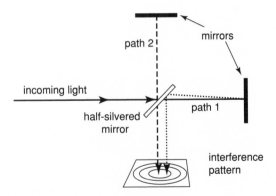

Figure 1.11. A simplified Michelson inferometer.

parts in a hundred million. In 1880, Michelson was granted a years' leave of absence from the navy to go to Germany and continue his research. While there, he invented his most important piece of scientific instrumentation—what came to be called the "Michelson interferometer." Figure 1.11 is a simplified diagram.

A beam of light enters the apparatus from the left. It encounters a half-silvered mirror that splits the beam. As shown, half the beam follows the path 1 and half path 2. Both beams encounter mirrors as shown and are reflected back and rejoin. Suppose that the time of transit for the two beams in their respective paths is somewhat different. This means that when they rejoin, the light waves will be slightly displaced from each other. A peak will no longer coincide with a peak, for example. The two beams will interfere and produce an interference pattern that will reflect this. This is how you can tell that there was a time difference in the paths. It is a very sensitive method and Michelson was sure that it could be adapted to measure the speed of the Earth in the aether. Before we discuss what actually happened when he tried this, we need to explain what the effect is. We have already done half the work. Suppose the Earth is moving in the direction of the path 1 beam. This is just the setup we discussed above. So we know that the round trip time which I will call τ_1 is $\tau_1 = \frac{2L}{c} \times \frac{1}{\left(1-\frac{v^2}{c^2}\right)} \cong \frac{2L}{c}(1 + \frac{v^2}{c^2})$. Finding τ_2 is a little more complicated. We are going to need the Pythagorean theorem from Euclidean geometry, which Einstein proved at the age of twelve and whose proof I give in the

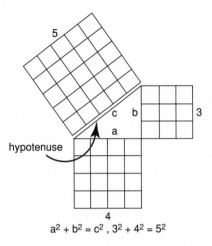

$$a^2 + b^2 = c^2 \; , \; 3^2 + 4^2 = 5^2$$

Figure 1.12. A familiar example of Euclidian geometry and the Pythagorean theorem.

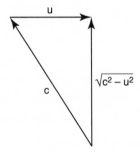

Figure 1.13. An example of the Pythagorean theorem at work.

next chapter. Here, I just want to remind you of what it says. Figure 1.12 may take you back to high school. The particular triangle with sides of magnitude 3, 4, and 5 may, or may not, recall fond memories.

Here is how I am going to use the theorem. Suppose, as Figure 1.13 indicates, the Earth is moving with a speed u in the direction shown. Light in the aether travels with the speed c. If I want to end up at the mirror I have to alter path 2 to take this into account. Michelson explained that this was like swimming with a stream flowing and trying to land on the bank just opposite where you started. You have to point upstream. The speed you have going in the direction you really want is the straight side of the right triangle in Figure 1.13.

The hypotenuse is c, and the other side is u, so by the Pythagorean theorem the speed you have going in the direction you want is given by $v = \sqrt{c^2 - u^2}$. If we factor out the c then the time it takes for the trip across and back is $\tau_2 = \frac{2L}{c} \times \frac{1}{\sqrt{1 - \frac{u^2}{c^2}}} \cong \frac{2L}{c}(1 + \frac{1}{2}\frac{u^2}{c^2})$.[6] The factor of one-half comes from the square root. In our case u is the speed v of the Earth through the aether. Thus for this case $\tau_1 - \tau_2 \cong \frac{L}{c} \times \frac{v^2}{c^2}$. It was this insanely small number that Michelson was proposing to measure or nearly. In his original analysis Michelson was off by a factor of two. He predicted an effect twice as large. But still, how can one measure it? I make use of the correct expression which Michelson soon adopted.

If we multiply the expression for the time difference by the velocity of light, c, we find the distance by which the two waves arriving back at the observer have been shifted. The question is what to take for L. I will shortly describe Michelson's most precise interferometer where L is approximately 10 meters—a thousand centimeters. It is not exactly this, but close enough so that we can keep the arithmetic simple. Putting in the numbers, we find an expected shift of about a hundred thousandth of a centimeter—10^{-5} centimeters. To what should we compare this? Michelson used yellow light from a sodium lamp. This light has a wave length of about 5.9×10^{-5} centimeters which means that the shift expected is a substantial fraction of this and should be observable. A problem may have occurred to the reader. It certainly occurred to Michelson. We have been assuming that the length L in the two arms is identical. To make the experiment work this statement must be accurate to a ten thousandth of a centimeter, or less. It was impossible for Michelson to guarantee this kind of accuracy. But he had a very clever idea. He would build the interferometer so that he could rotate the arms. If he rotated by ninety degrees this would interchange the role of the two arms. If he averaged the two readings, the difference in the path lengths would drop out and any shift in the interference bands would be the effect he was expecting to find. Thus Michelson was now prepared to build his interferometer and to measure the speed of the Earth through the aether.

---

[6]We have used the small x expansion $\frac{1}{\sqrt{1-x}} \cong 1 + \frac{1}{2}x$.

The work was carried out in the winter and spring of 1880 to 1881. The interferometer, which incidentally was partially financed by Alexander Graham Bell, was located on the stone base of circular room below the central tower of an observatory outside Berlin. This guaranteed there would be no unwanted vibrations. Michelson then began to observe. He found nothing. There was no shift at all. After six months of observation he summarized what he had found, or what he had not found.

> The interpretation of these results is that there is no displacement of the interference bands. The result of the hypothesis of a stationary ether [Michelson's spelling] is thus shown to be incorrect, and the necessary conclusion follows that the hypothesis is erroneous.
>
> The conclusion directly contradicts the explanation of the phenomenon of aberration which has been hitherto generally accepted, and which presupposes that the earth moves through the ether.

This latter is a reference to an observation made in 1725 by the British astronomer James Bradley. A homey analogy to what Bradley observed is a common experience of bicycle riders riding in the rain. If you are at rest the rain comes straight down. If you are moving the same rain comes at you from an angle, a result of compounding your velocity with that of the rain drops. You might be led to think that the source of the rain has shifted. Bradley observed a seasonal shift in the positions of the stars that reflected the Earth's motion, which he attributed to the fact that in the stationary aether the light would come straight down to the Earth, if it was not in motion. It would come in at an angle if the Earth was in motion and the stars would appear displaced. Bradley used a particle theory of light to describe what he saw. When we discuss relativity we will use the wave theory which makes the description a little more complicated. From the aether theory point of view this experiment showed that the Earth is moving through the aether with little or no drag. If the Earth dragged the aether along than the starlight would have come straight down.

Michelson felt obliged to explain to his patron Bell what had happened—the negative result of his experiment. In his letter to Bell he offered his own explanation which seemed to ignore Bradley's experiment. It was a revival of an earlier idea of some of the aetherists that the aether, in the neighborhood of the Earth, might at least be partially dragged by it. This notion was largely abandoned because it seemed to lead to contradictions. But, in desperation, Michelson clutched at this straw.

In fairness to Michelson, there was some experimental evidence that apparently suggested this. Following some earlier work by the French physicist Augustin Fresnel, in 1851, the French physicist Hippolyte Fizeau measured what became referred to as the "Fresnel dragging coefficient." It was well known that light in a medium like water is slowed. A measure of this is the so-called "index of refraction", n, which in the simplest wave theory case, is just the ratio of the speed of light, c, to the speed in the medium, v. In these cases n is greater than one. For water, for example, it is 1.33. The question Fresnel raised is, suppose the water is moving. Three possibilities presented themselves. If the water does not drag the aether there is no effect. If it totally drags the aether there is a maximum effect and if it partially drags the aether then there is a partial effect. Fresnel proposed a formula which, as far as I can see, was an educated guess. He said that the combined effects of the dragging and the slowing down in the medium would produce an effective light velocity c' given by $c' = v + (1 - 1/n^2)w$, where w is the speed of the water and $1 - 1/n^2$ is the Fresnel dragging coefficient. This is what Fizeau measured and found it to be about 0.44. Thus this experiment seemed to show that the aether was partially dragged, while the Bradley experiment seemed to show that it was not dragged at all. One of the things that the pre-Einsteinian physicists tried to do, with considerable difficulty, was to derive Fresnel's formula. In relativity, as we shall see, one can derive it in a single line. The confusion that these 19th-century results engendered is eliminated. A hint, which was ignored, is that in this example, the Newtonian addition of velocities has evidently broken down. Naively one might think that c' should be just the sum of the two speeds v and w. It is not, and this, as we shall see in the next chapter, is a feature of Einstein's relativity.

I am now going to go outside the chronology. I want to get Michelson off the stage so I can bring in the last actors in this pre-Einstein drama—the theorists. At the end of his German stay, Michelson decided to leave the navy so he could concentrate completely on his scientific work. He was offered a position at the recently founded Case School of Applied Science—now Case Western Reserve university—in Cleveland, Ohio. For a few years he did not make any further attempts to measure the Earth's velocity in the aether. But, in 1884, he was invited to attend a conference in Montreal. Another attendee was the chemist Edward Morley who was at Western Reserve University, also in Cleveland. Morley, who was fourteen years older than Michelson, had studied astronomy, and was interested in Michelson's work. They decided to join forces and began the construction of the mother of all interferometers shown below (Figure 1.14).

Because of a fire at Case which destroyed much of his laboratory, Michelson relocated the work to Western Reserve. He and Morley made several improvements. Among them was floating the stone on which the instrument lay, on mercury. This reduced vibrations and made it easy

Figure 1.14. Michelson and Morley's 1887 inferometer built in the basement of Western Reserve. (Photo courtesy of Case Western Reserve Archive)

to turn. The arms were considerably longer and sixteen mirrors were used rather than two. You can see the extra mirrors in the photograph. These multiple reflections extended the path length of the light to twenty two meters–nearly seventy feet. In July of 1887, they performed what is known as the "Michelson-Morley" experiment. In fact, people who have not studied this history, often write as if this was the entire enterprise– ignoring what Michelson had done earlier. After two days of observation they confirmed that there was no effect–which Michelson had already found earlier with less accuracy. Michelson seemed to regard all of this as a failure. It is not what he had expected, and he never seemed fully to reconcile himself to it, or to the developments that followed. In 1889, he left Case ending up eventually at the University of Chicago in 1892, where he was a member of the faculty until his death in 1931. Now to the theorists.

There are three I want to discuss. The first is the Irish physicist George Francis Fitzgerald. Fitzgerald had done a good deal of post-Maxwellian electrodynamics. Apart from the historians of science, this work is essentially forgotten. What is not forgotten is a **one paragraph** note he published in the American journal *Science* in 1889. Here is what he wrote,

I have read with much interest Messrs. Michelson and Morley's wonderfully delicate experiment attempting to decide the important question as to how far the ether is carried along by the earth. [This is an odd view of what Michelson was trying to do.] Their result seems opposed to other experiments showing that the ether in the air can be carried along only to an inappreciable extent. I would suggest that almost the only hypothesis that can reconcile this opposition is that the length of material bodies changes according as they are moving through the ether or across it, by an amount depending on the square of the ratio of their velocities to that of light. We know the electric forces are affected by the motion of the electrified bodies relative to the ether, and it seems a not improbable supposition that the molecular forces are affected by the motion, and that the size of a body alters consequently. It would be very important if secular experiments on electrical attractions between permanently electrified bodies, such as in a very delicate quadrant electrometer, were instituted in some of the equatorial parts of the earth to observe whether there is any diurnal and annual variation of attraction—diurnal due to the rotation of the earth being added and subtracted from its orbital velocity, and annual similarly for its orbital velocity and the motion of the solar system.

This is the entire the paragraph. I want to focus on the part about the length of material bodies changing. What does Fitzgerald mean? Let us go back to the two paths. The time it takes for the light to go to travel back and forth on the path 2 is $\tau_2 = \frac{2L}{c} \times \frac{1}{\sqrt{1-\frac{v^2}{c^2}}}$ while for the other path is $\tau_1 = \frac{2L}{c} \times \frac{1}{1-\frac{v^2}{c^2}}$. It is on the path 1 where the arm is in the direction of motion. What Fitzgerald is saying is that if the length L in the direction of motion contracts by a factor of $\sqrt{1-\frac{v^2}{c^2}}$, then the two times become the same and Michelson's result is explained. I do not want to get into Fitzgerald's suggested possible explanation for this bizarre behavior. I want to ask if it contradicts our common experience. We do not observe things contracting when they move, but our motions are infinitesimally slow compared to the speed of light. It is instructive to make the following little estimate. The mean radius of the Earth is given as 251,106,299 inches, while v^2/c^2 is about one hundred millionth. Thus the contraction that Fitzgerald was asking for would shrink the radius of the earth by less than three inches!

His suggestion became known to several physicists, among them the Dutch physicist Hendrick Antoon Lorentz, who had had the same idea independently. Of the physicists working in electromagnetism between Maxwell and Einstein, Lorentz was the greatest. Einstein once noted that of all the physicists he had met, Lorentz impressed him the most. Lorentz was born in 1853 in Arnhem. He entered the University of Leyden in 1870. He obtained his doctors degree at age twenty-two and three years later he was appointed to the Chair of Theoretical Physics which had been newly created for him. He remained at the university, despite many offers to go elsewhere, for the rest of his life. He died in 1928. Beginning with his PhD thesis he was occupied during his whole long career with electromagnetism. We are only going to focus on a small part.

In 1897, the British physicist J.J. Thomson, who was studying electrical discharges in a high vacuum glass tube, discovered that a stream of particles was passing from the negatively charged to the positively charged metal plates in the tube. He was able to deflect the particles with electromagnetic fields and found that they were negatively charged and had a smaller mass than any known atom. He had discovered what came to be called the "electron." What Lorentz did was to derive from the Maxwell equations the force law that such an electron would obey in electric and magnetic fields. We still use it today. The electric field part is rather intuitive. The force exerted on the charge is proportional to the field and in the direction of the field. The magnetic field part isn't at all intuitive. It depends on the velocity of the electron and the field. When the electron is at rest, the magnetic force vanishes. Moreover, the magnetic force is directed at right angles to the field and velocity. We usually teach this with the so-called "right-hand rule" as illustrated in Figure 1.15. If the index finger is in the direction of the velocity and the middle finger is in the direction of the field, then the thumb is in the direction of the force. This is for a positive charge. For a negative charge like the electron the force would be in the opposite direction.

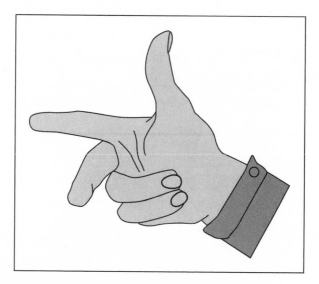

Figure 1.15. The right hand rule.

This force violates the Galilean relativity principle discussed earlier in the chapter. We argued that Newton's laws obey this principle because the forces that he considered, such as gravitation, do not depend on velocities. This force does. What this tells us is that if we insist on a relativity principle for electromagnetism, it will involve something new.

Lorentz was a believer in a stationary aether. He wanted to reconcile this with the Michelson result. To this end he wanted to show explicitly how the contraction could come about. He had an electromagnetic model of molecules in mind, meaning that molecules were held together by electromagnetic forces. His idea was to study how these forces would be modified if the molecule was set in motion. He only considered uniform motions with constant speeds much smaller than the speed of light. He needed to study the Maxwell equations in this moving system. In the course of doing this he discovered something very important. Figure 1.16 shows a rest system and a moving system. The prime coordinates refer to the moving system. If a point in the rest system has a coordinate x, then the same point in the moving system will have a coordinate x'. If we relate this to the x coordinate we see from the diagram that $x' = x - vt$, where we are assuming that the prime system is moving to the right with a speed v. This is the Galilean relativity transformation for the coordinate. This will not leave the Maxwell equations in the same form they have in the rest system. In other words, this transformation does not satisfy the relativity principle for electrodynamics. Lorentz did not put it this way. For purposes of his analysis, he wanted to have the Maxwell equations the same in the moving system because then he could apply what he

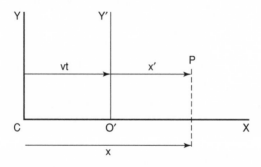

Figure 1.16. A rest system and a moving system.

knew about the equations in the rest system. To accomplish this Lorentz found that he had to transform the **time!**

Lorentz was considering slow speeds so he neglected terms of higher order in v/c. With this assumption he found that the following time transformation worked, $t' = t - \frac{v}{c^2}x$. If he included this with the transformation $x' = x - vt$, the Maxwell equations for slow speeds retained their form. Lorentz called t' the "local time," presumably because it varied from point to point in space. For Lorentz the local time was an artifact–a trick to make his calculations simpler. He never considered the possibility that this local time might have some physical reality as, say, a time measured by actual clocks in the moving system. He published this transformation in 1895, in a book, *Versuch einer Theorie der elektrischen und optischen Erscheinungen in bewegten Körpern*, which was published in Leyden. It is quite certain that Einstein read this book. It is equally certain that he did not read any of the rest of Lorentz's papers published before 1905. They were all published in local Dutch journals to which Einstein had no access.

Historians of science disagree on the extent that Lorentz anticipated Einstein. Some say a great deal and some say rather little. I want to consider the last paper Lorentz wrote before the revolution of 1905. It was published in 1904. There was an English version which appeared in the Academy of Sciences of Amsterdam. The title is "Electromagnetic Phenomena in a System Moving with any Velocity Less than that of Light." The paper deals with the transformation of the Maxwell equations when the speeds are not necessarily small. There is a generalization of the time transformation for this situation. In all honesty I find this generalization a bit opaque. I cannot readily connect it with the transformation that Einstein was soon to derive. Lorentz discusses a number of phenomena including the contraction of lengths. What struck me is what he assumed. He states explicitly, —the italics are in the paper—"I shall now suppose *that the electrons, which I take to be spheres of radius R in the state of rest, have their dimensions changed by the effect of a translation, the dimensions in the direction of motion becoming βl times and those in the perpendicular direction l times smaller.*" Here $\beta = \sqrt{1 - \frac{v^2}{c^2}}$. The quantity "l" is some sort of scale factor.

One might well argue, on the basis of common sense, that there is no reason why there should be a contraction in the direction perpendicular to the speed, so that 'l' should be set equal to one. Why Lorentz includes this factor I do not know. But note, he **assumes** the electron contraction. From this he derives the contraction of the molecule, but it is all tied to this electromagnetic model and not to general principles. Here is where his work vividly contrasts with that of Einstein. As we shall see, Einstein's relativity does not depend on any specific molecular model. It depends on a profound analysis of space and time. In this respect I would like to emphasize the following. Lorentz devised his transformations so that they would produce equivalent forms of the Maxwell equations in the rest system and in any system moving uniformly with respect to it. But the same transformations do **not** produce equivalent forms for Newton's law! Put graphically, the Lorentz transformations preserve the relativity principle for electromagnetism at the cost of not preserving it for Newtonian mechanics. This did not seem to concern Lorentz. For him the transformations were a mathematical artifact. The "local time" was not something you would measure with a clock. It was just a change of variables. Hence he never considered the possibility that to make the relativity principle universal you would need a new form of mechanics. The pre-Einstein figure who did seem to understand this was the French polymathic genius Henri Poincaré to whom we now turn.

In our voyage through this history we have encountered several geniuses. I would put Poincaré near the top. He seems to have had knowledge of all the sciences and to have made profound contributions to several, although he was nominally a mathematician. He was also a wonderful stylist and wrote several books on the philosophy of science, as well as more technical monographs. As far as I am concerned, he was the one person before Einstein to really understand the issues. Poincaré was born in 1854, in Nancy, France. He died in 1912, so that he was able to witness the relativity revolution and to contribute to it. He spent most of his career at the University of Paris and by the end of it he had written some five hundred papers. I want to present three quotations from Poincaré which I think tell the story. In 1899, in a lecture at the

Sorbonne Poincaré stated that "I regard it as very probable that optical phenomena depend only on *relative* motions of the material bodies, luminous sources, and optical apparatus concerned, and that this is true not merely as far as quantities of the order of the squares of the aberrations [v^2/c^2] but *rigorously*." In short, in 1899, Poincaré is clearly stating that what the Michelson experiment, and the others like it, show is that the principle of relativity applies to these electromagnetic phenomena. According to this principle no effect was to be expected in the Michelson experiments, and none was found. In 1900, Poincaré addressed an International Congress of Physics in Paris and asked, "Our aether, does it really exist? I do not believe that more precise observations could ever reveal anything more than *relative* displacements." Perhaps, he seems to be saying, the entire issue of the aether is a chimera. Finally, in an international congress held in St. Louis in 1904, he states, "According to the Principle of Relativity the laws of physical phenomena must be the same for a "fixed" observer as for an observer who has a uniform motion of translation relative to him: so that we have not, and cannot possibly have any means of discerning whether we are, or are not carried along in such a motion." He must have had a clear idea that the relativity of Newtonian mechanics and this generalized relativity were incompatible, for he adds, "From all these results there must arise an entirely new kind of dynamics, *which will be characterized above all by the rule that no velocity can exceed the velocity of light.*" We can be sure that Einstein did not read this statement and we can also be sure that Poincaré did not know that there was a twenty-six year old patent examiner in Bern who would create this new dynamics. This is the subject of the next chapter.

2
Einstein's Theory of Relativity

In the autumn of the same year, in the same volume of the *Annalen der Physik* as his paper on Brownian motion, Einstein published a paper which set forth the relativity theory of Poincaré and Lorentz with some amplifications, and which attracted much attention.
— Sir Edmund Whittaker

Was Herr Einstein hat gesagt ist nicht so blöde (What Mister Einstein has said is not so stupid.)
— The very young Wolfgang Pauli after an Einstein lecture

↪ PROLOGUE

True enough, physics was also divided into separate fields, each of which was capable of devouring a short lifetime of work without having satisfied the hunger for deeper knowledge. The mass of insufficiently connected experimental data was overwhelming here also. In this field, however, I soon learned to scent out that which was able to lead to fundamentals and to turn aside from everything else, from the multitude of things which clutter up the mind and divert it from the essential.

—Albert Einstein

In my sophomore year I took two more courses with Professor Frank. The fall course was another lecture course that was more philosophically oriented than the first I had taken. In the spring, I followed with a reading course, just the two of us. Together, we read people like Mach, Poincaré, and Wittgenstein. We also talked a great deal. He told me about how he, and a like-minded group of young people in Vienna, began a philosophical movement. They became known as the "Vienna Circle" and the movement, "Logical Positivism." The main idea was to seek a unity of knowledge, physics, mathematics, and economics, which would be free of what they considered to be extraneous metaphysics. They met from time to time in various coffee houses in Vienna—establishments that were tolerant of much discussion and relatively little consumption of coffee. When I visited Vienna many years ago an elderly waiter in one of the coffee houses claimed to have remembered them. The only equivalent Professor Frank could find in Cambridge was a Hayes Bickford cafeteria on Harvard Square. We often went there, and when we did, I asked Professor Frank questions about Einstein himself, as well as his physics. One question I asked may seem somewhat crazy. I wanted to know if Einstein when young would have seemed smart. By this time I had met several very smart people, some of whom seemed smart and some did not. Years later I had the chance to meet and observe Niels Bohr who, after Einstein, was probably the greatest physicist of the twentieth century. He certainly did not seem smart. He was very slow of speech and often incomprehensible. I wanted to know where Einstein would have fitted in this spectrum. Professor Frank told me that the young Einstein seemed very smart indeed. He was frequently given to, what Professor Frank in his inimitable accent, called "krecks." These "krecks" often got Einstein in trouble. People felt that he was a little too smart for his own good.

By the end of that Spring I had taken an introductory course in physics which I found incredibly dull after what I had learned in Professor Frank's courses. I had also taken a course in calculus. I then had the lunatic idea that it would be nice to talk to Einstein. About what, I cannot imagine. Professor Frank must have put a word in for me because a letter arrived on June 3,1949, from Princeton. It was in English, and typed, so I imagine that Helen Dukas must have typed it. It reads,

Dear Mr. Bernstein

I am sending you enclosed paper in which I expressed opinions from an epistological [sic] point of view. I do not give oral interviews to avoid misinterpretation.

Sincerely yours,
A. Einstein

The paper in question, which I still have, was the Herbert Spencer lecture Einstein gave at Oxford in June of 1933. He was, it turned out, on his way to the United States and would never return to Europe. The lecture was called "On the Method of Theoretical Physics" It was really a description of *his* method, or what had become his method after his discovery of general relativity with its new theory of gravitation. We will discuss this briefly at the end of this chapter, but let me quote some of what Einstein said,

Our experience hitherto justifies us in believing that nature is the realization of the simplest conceivable mathematical ideas. I am convinced that we can discover, by means of purely mathematical constructions, these concepts and those lawful connections between them which furnish the key to the understanding of natural phenomena. Experience may suggest the appropriate mathematical concepts, but they most certainly cannot be deduced from it. Experience, remains of course, the sole criterion of physical utility of a mathematical construction. But the creative principle resides in the mathematics. I hold it true that pure thought can grasp reality, as the ancients dreamed.

The general theory of relativity is the only example I know of the creation of a profound theory of physics that has been successfully constructed this way. There was essentially no experimental input. Einstein was guided by criteria of symmetry and simplicity. It obviously made an immense impression on him. It became his working method for the latter part of his life when he was trying to construct a theory of everything. As far as I can see, this attempt was a total failure, but he kept working on it almost to his last breath.

After I graduated, I kept close contact with Professor Frank. When he retired from Harvard I helped clean out his office. He had a roll top desk with letters–many unopened–along with other miscellanea stuffed in various pigeon holes. We opened a few of the letters. There was one from Erwin Schrödinger, one of the architects of the quantum theory. It was in German and Professor Frank translated. The first sentence began "Just between us daughters of parsons . . . " and was a complaint about Einstein, with whom Schrödinger was having some sort of dispute. There was also a very dusty scroll. We opened that. It was an etching of Einstein done in 1932. The artist had signed it along with Einstein. Professor Frank asked me if I wanted it. It hangs on a wall in my apartment.

We are now going to begin our exposition of Einstein's relativity. But before doing so, there is some unfinished business. I promised to show you Einstein's proof of the Pythagorean theorem. We are going to use the theorem again, so I will now show you his proof. If you are happy to accept the theorem, you can skip over this section.

↬ A Pythagorean Interlude

At the age of 12 I experienced a second wonder of a totally different nature [The first "wonder" occurred when he was four or five and his father gave him a compass.]; in a little book dealing with Euclidean plane geometry, which came into my hands at the beginning of a school year. Here were assertions, as for example, the intersections of the three altitudes of a triangle in one point, which–though by no means evident–could nevertheless be proved with such certainty that any doubt appeared to be out of the question. This lucidity and certainty made an indescribable impression on me. That the axiom had to be accepted unproved did not disturb me. In any case it was quite sufficient for me if I could peg proofs upon propositions the validity of which did not seem to me to be dubious. I remember than an uncle told me the Pythagorean theorem before the holy geometry booklet came into my hands. After much effort I succeeded in "proving" this theorem on the basis of the similarity of triangles; in doing so it seemed "evident" that the relations of the sides of a right-angled triangles would have to be completely determined by one of the acute angles. . . .

To see what Einstein meant consider the similar right triangles in Figure 2.1.

"Similar" means that all the sides of the triangles have been scaled up, or down, by the same scale factor so the angles remain the same. Let me call the length from A to C, "a" and the length from A$'$ to C$'$, "a'". Let me call the scale factor "l". Then $a = a'l$, and a similar relation holds for the other sides with the **same** l. If we call the length from C to B, "c" and the length from C$'$ to B$'$ "c'," then from our assumption we have $a/c = a'/c'$, because the scale l cancels out. We can rewrite this as $c'/c = a'/a = b'/b$, where the "bs" are the hypotenuses of the triangles. This will be the main tool we will need in proving the Pythagorean theorem. We are now going to consider the triangle below. Before I explain how, let me revise the notation for the lengths of the sides. Call

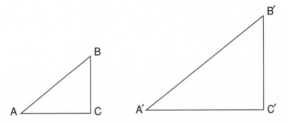

Figure 2.1. Similar right triangles.

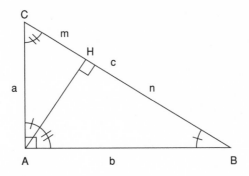

Figure 2.2. Construction of two triangles from one.

the length from A to C, "*a*," the length from C to H, "*m*," the length H to B, "*n*." The length C to B, I call "*c*" and, finally, the length B to A, "*b*." Note that $m + n = c$. I have constructed two triangles by dropping a perpendicular line from A to the hypotenuse of the large triangle (Figure 2.2). The two small triangles are similar to each other, and to the large triangle. The way to see this is to take the angles at A arbitrary, but adding up to 90°, and then showing that they must be as drawn.

We can now finish off the argument. We have the similarity relations $(m + n)/b = b/n$ or $b^2 = n(m + n) = nc$ and $(m + n)/a = a/m$ or $a^2 = m(m + n)$. Adding these together we have $a^2 + b^2 = c^2$, which is the Pythagorean theorem.

Despite the appeal of the "holy geometry book," Einstein never had much interest in pure mathematics for its own sake. Towards the end of his life he wrote,

> I saw that mathematics was split up into numerous specialties, each of which could easily absorb the short lifetime granted to us. Consequently I saw myself in the position of Buridan's ass which was unable to decide upon any specific bundle of hay. This was obviously due to the fact that my intuition was not strong enough in the field of mathematics in order to differentiate clearly the fundamentally important, that which is really basic, from the rest of the more or less dispensable erudition. Beyond this, however, my interest in the knowledge of nature was also unqualifiedly stronger, and it was not clear to me as a student that the approach to a more profound knowledge of the basic principles of physics is tied up

with the most intricate mathematical methods. This dawned upon me only gradually after years of independent scientific work.

I think it is fair to say that the mathematics used in the "Miracle Year" papers does not go beyond what a physics major would learn in the first few years of study. The "mathematization" of the special theory of relativity–the beautiful four-dimensional formulation presented in 1908, by Hermann Minkowski, which we all now use, and which I will touch on later in the chapter, caused Einstein to say that once the mathematicians had gotten a hold of his theory he no longer understood it himself–one of his "krecks." Minkowski had been one of Einstein's teachers at the Poly in Zurich. When he discovered that it was Einstein who had created the theory he was amazed that, that "lazy dog" could have done something like this. Einstein cut many of Minkowski's classes, which did not sit well. When Einstein began working seriously on gravitation he had to generalize what Minkowski had done, creating some new mathematics along the way. With this prologue we can now begin our study of the special theory of relativity.

∽ Space and Time

> If, for instance, I say, "That the train arrives here at 7 o'clock," I mean something like this; "The pointing of the small hand of my watch to 7 and the arrival of the train are simultaneous events."
> —Albert Einstein

I have often asked myself what I would ask Einstein if I could meet him now, now that I know so much more. There are many questions that come to mind but one persistent one is, "What took you so long?" Here is what I have in mind. In 1949, Einstein published his "Autobiographical Notes" from which I have been quoting. They are not very autobiographical if you are looking for what Einstein called somewhat disdainfully the "merely personal." You will find nothing about his wives and children for example. It is a scientific autobiography in which he tries to explain how his ideas evolved. When it comes to relativity he writes the following,

After ten years of reflection such a principle [relativity] resulted from a paradox upon which I had already hit at the age of sixteen: if I pursue a beam of light with the velocity c (velocity of light in a vacuum), I should observe such a beam of light as a spatially oscillatory electromagnetic field at rest. However, there seems to be no such thing according to Maxwell's equations. From the beginning it appeared intuitively clear to me that, judged from the standpoint of such an observer, everything would happen according to the same laws as for an observer who, relative to the earth, was at rest. For how otherwise, should the first observer know, i.e., be able to determine that he is in a state of fast uniform motion?[1]

Let me deconstruct this before I go on to quote the next paragraph. In the first place, there is no evidence that Einstein knew much about the Maxwell equations at the age of sixteen. There were no courses in Maxwell's electrodynamics taught at the Poly when Einstein was there and which he entered at the age of seventeen. He began to learn this material on his own a few years after he had matriculated. Whatever he concluded from the Maxwell equations, must have come later. In the second place, we have some idea of the physics Einstein was pursuing around the age of sixteen. There is a document dated 1895 entitled. "On the Investigation of the State of the Ether in a Magnetic Field." Einstein was then an aether physicist and remained so until at least 1901. That year he wrote a letter to a friend that he had thought of a new experiment to measure the motion of the Earth through the aether. There has been much discussion of whether Einstein had heard of the Michelson-Morley experiment. He said various things at various times. He noted, for example, that if he had known of it he would have mentioned it in his relativity paper. But his paper does not contain references to any other paper or monograph, including the 1895, one of Lorentz which we know he studied. Late in his life he said that he could no longer remember if

---

[1] Schilpp, *op.cit.*, p. 53. My understanding of what Einstein meant here is that if you could catch up with a light wave it would appear "frozen." It would be like coming across a pond which had a wavy surface but the waves did not move. This would certainly appear "paradoxical."

he had heard of it, but in any event, it would have played only a minor role, if any, in his invention of the theory. The reason is, as we have already seen in the last chapter, that what the Michelson-Morley experiment shows is that the relativity principle applies to electromagnetic phenomena—something that Einstein was convinced of anyway. So as to why it took Einstein so long, one part of the answer, it seems to me, is that he started much later than the age of sixteen and, indeed, as I noted, was an aether theorist until at least 1901. Another part of the answer is hinted at in the paragraph that follows the one I quoted:

> One sees that in this paradox the germ of the special theory of relativity is already contained. Today everyone knows, of course, that all attempts to clarify this paradox satisfactorily were condemned to failure as long as the axiom of the absolute character of time, viz., of simultaneity unrecognizably was anchored in the unconscious. Clearly to recognize this axiom and its arbitrary character really already implies the solution of the problem. The type of critical reasoning which was required for the discovery of this central point was decisively furthered in my case by the reading of David Hume's [the 18th-century Scottish philosopher] and Ernst Mach's philosophical writings.

There is a lot to digest here. In the first place, what "paradox" are we talking about? The following three things are incompatible; Newtonian mechanics, Maxwell's electromagnetic theory, and the relativity principle extended to cover electromagnetism. This is the basic content of the Michelson experiment. A paradox would arise if you insisted that all three of these things be simultaneously true. To understand the rest of the paragraph, let us ask ourselves how we got into this position. The culprit is the common sense Newtonian addition theorem for velocities. This is what persuaded Michelson that he could detect the Earth's motion through the aether. The speed of light as observed on the moving Earth was supposedly affected by the Earth's orbital velocity—so claimed the addition theorem. We must then ask, how is this theorem derived? Below I have a diagram that is also in the last chapter. In it you will see two coordinate systems—one in motion uniformly to the right with respect to the other. Incidentally, Professor Frank told me a funny story. He went to visit Einstein in Berlin just after Einstein had published his popular exposition of relativity: *"Relativity, the Special and the General Theory; A Popular Exposition."* Einstein was by then remarried to his cousin Elsa and was living with his two stepdaughters. Einstein explained that his book was so clear that his young stepdaughters understood it all. When Einstein left the room, Professor Frank asked one of them if it was true that she really understood the book. "Yes," she said, "Everything except what is a coordinate system." Anyway, below we have two of them.

Let us as a matter of convention call the X, Y system the "rest system" and the x', y' system the "moving system." We could carry out the same analysis with the roles reversed with the moving system now going backward. In the rest system the coordinate of the point of interest is x. We will not bother with the y-coordinates. In the moving system the coordinate of the same point is x'. In a time, t, the moving system has moved a distance vt, if v is the relative velocity. So we must subtract this distance to find x'. Thus $x' = x - vt$. This supposes that the point x is at rest. But let us suppose it is in motion with respect to the rest system with a speed u. Thus $x = ut$. In Figure 2.3, I have assumed that u is greater than v, so the point x has outrun the origin of the moving

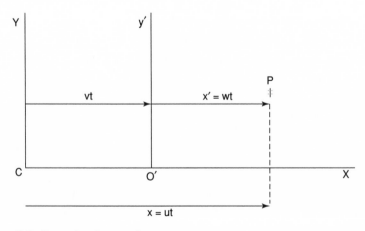

Figure 2.3. Example of a coordinate systems.

system. As I have written the equations, they both start off at the same place at $t = 0$. Thus $x' = t(u - v)$. But the distance traveled by the x', y' origin with respect to the moving point x is $wt = t(u - v)$, where w is the speed of the point p as viewed in the x', y' system. Now if we cancel the ts we have the addition theorem $w = u - v$ that has gotten us into all the trouble. What can possibly be wrong with this derivation? This brings us to the crucial point.

In this derivation we have assumed that time as measured in the moving system is the **same** as time measured in the X, Y system; the "rest" system. We have bought the Newtonian notion of absolute time. At some point, Einstein must have realized that this was the crux of the matter. When this happened, I am not sure. I think it happened not long before he wrote his relativity paper and I think it was clarified in the course of conversations that he had with his friend and patent office colleague Michele Besso. The only acknowledgment to anyone in Einstein's relativity paper is what he writes at the end; "In conclusion I wish to say that in working at the problem dealt with I have had the loyal assistance of my friend and colleague M. Besso, and that I am indebted to him for several valuable suggestions." A few words about Besso.

Besso was born in Zurich in 1873, making him some six years older than Einstein. He was a precocious child who, like Einstein, acquired in the Gymnasium a reputation for disrespect. He complained about the quality of the mathematics instruction. Like Einstein, he was educated at the "Poly." The two men seem to have met at a musical soirée and immediately struck up a friendship that lasted their entire lives. Besso died a month before Einstein in 1955. Besso was more of an engineer than a physicist. As brilliant as he was, he had difficulty focusing. He had a sister named Bice Margherita Louisa Besso who married a Florentine count named Rusconi. They had a daughter Laura who married the *New Yorker* writer Niccolò Tucci. In 1947, he went with his mother-in-law to Princeton to visit Einstein. He wrote this visit up and recorded an exchange between Bice and Einstein,

"Herr Professor," she asked, in German (the whole conversation, in fact, was in German), "this I really meant to ask you for a long time—why hasn't Michele made some important discovery in mathematics?"

"*Aber*, Frau Bice," said Einstein laughing, "this is a very good sign. Michele is a humanist, a universal spirit, too interested in too many things to become a monomaniac. Only a monomaniac gets what we commonly refer to as *results*."

Besso, with his lack of ego involvement, and his broadness of intellect, must have made a perfect sounding board for Einstein's new ideas. What then is the point? Einstein's observation was that the specification of the time of an event involves, in fact, the specification of two simultaneous events. First, one must specify the event—the arrival of a train "here," to use Einstein's example, and some marker, such as a pointer, on the dial of a clock. For the moment I will not worry about what a "clock" is. Shortly I will give you an example which illustrates the issues. A Newtonian, who had not reflected deeply on the matter, would say that the "simultaneity" of events has an absolute significance, so time is universal. But is it?

Figure 2.4 shows Einstein's favorite example–trains. There is a train moving to the right with a speed v. With respect to an observer on the ground, lightning bolts strike at points A and B simultaneously. How do we know? We can station ourselves in the middle and note that the light comes to us from the two sides simultaneously. In this discussion we always assume that we can tell whether two events are simultaneous when they occur at the same point in space. We infer from this that the lightning bolts have struck simultaneously in our resting frame some time in the past. But what about the moving train? What would an observer in the middle of the train claim? I will give the answer that a Newtonian might have given and then later, when we have done some relativity, I will give the Einsteinian answer. What a Newtonian might have said is that to the train observer the events will not appear to have been simultaneous. There are two effects. The Newtonian would say the speed of light is different in the two directions; $c + v$, in one, and $c - v$ in the

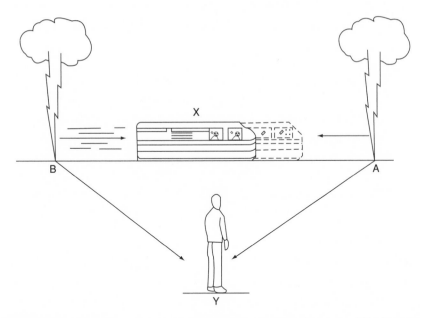

Figure 2.4. In Einstein's train example, the light from A will arrive at X before that from B. Hence X will observe the lightning at A as happening before that at B. Y, however, will observe the bolts of lightning to be simultaneous. This is an example of how observations from reference frames moving relative to each other reveal a different timing of events.

other. Moreover, the distances covered are different; shorter on one side than the other. Thus the Newtonian who used the same operational definition simultaneity of these distant events — the light arriving at the same time at the center — should have said that simultaneity is relative and should have examined what this meant about the nature of time. Instead, the Newtonian insisted on the use of "absolute" time which essentially by definition makes these events–the striking of the two distant lightning bolts — simultaneous for both observers. Whether Einstein passed through this Newtonian stage or whether he leapt right into relativity, I do not know. But it is time that we take the leap.

In the introduction to his paper Einstein spells out his program. He writes,

> ... the unsuccessful attempts to discover any motion of the earth relative to the "light medium," suggest that the phenomena of electrodynamics as well as of mechanics possess no properties

corresponding to the idea of absolute rest. They suggest rather that, as has already been shown to first order in small quantities, the same laws of electrodynamics and optics will be valid for all frames of reference for which the equations of mechanics hold good. We will raise this conjecture (the purport of which will hereafter be called the "Principle of Relativity") to the status of a postulate, and also introduce another postulate, which is only apparently irreconcilable with the former, namely that light is always propagated in empty space with a definite velocity c^2 which is independent of the state of motion of the emitting body. These two postulates suffice for the attainment of a simple and consistent theory of the electrodynamics of moving bodies based on Maxwell's theory for stationary bodies. The introduction of a "luminiferous ether" will prove to be superfluous inasmuch as the view here to be developed will not require an "absolutely stationary space" provided with special properties. . . .

In deconstructing this remarkable paragraph several things strike one. There is the sentence about the "luminiferous ether," which is eliminated by the stroke of a pen. A century's worth of work by many very distinguished physicists is declared "superfluous" by a twenty-six year old patent clerk. As you might imagine, this did not go down well. Some, like Michelson, never really accepted relativity. It took Lorentz several years. Entrenched ideas die hard. We have said enough so far, so that Einstein's formulation of the relativity principle should seem plausible. I would just like to adumbrate the parts about "unsuccessful attempts to detect any motion of the earth," and the reference to "first order in small quantities."

Einstein does not tell us which "unsuccessful attempts" he is referring to. It could be the Michelson experiment, but it could also be to an experiment that was begun in 1901 by the British physicist Frederick

---

[2]As I mentioned earlier, in the original paper Einstein uses "V" for the speed of light. "c" is now universally used so I will also use it. Only Einstein's German original uses "V." He also uses "V" in his short paper in which he derives $E = mc^2$ which he calls mV^2. The symbol "c" comes from the Latin *celeritas* — "swiftness."

Thomas Trouton. The final version done in collaboration with H.R. Noble was published in 1904. It did not use light but a different electromagnetic property.[3] This experiment should have been capable of detecting v^2/c^2 effects and didn't. We do not know what Einstein had in mind. What about the reference to "first order in small quantities?" As I have indicated, the "paradox" in this discussion arises if we try to maintain both the validity of the Newtonian form of the relativity principle, with its absolute time, and the Maxwell equations. However, this only manifests itself when we consider effects of order v^2/c^2. So long as we restrict ourselves to effects of order v/c there is no contradiction. Let me take a specific example that is interesting in its own right—the Doppler shift for light. We can envision two situations; the emitter of the light is moving towards the observer, or the observer of the light is moving towards the emitter. Common sense, let alone the relativity principle, should tell us that these two situations are interchangeable, that the only thing that should matter is our relative velocity. But this is not what happens in the aether theory. If the emitter is moving towards the observer one readily shows that the Doppler shift in wave length is proportional to $\frac{1}{1-v/c} \cong 1 + v/c$, while, if the observer is moving towards the emitter, the shift is simply $1 + v/c$. Thus, to order v/c, the relativity principle is preserved. This is generally true about all these aether theory electromagnetic and optical phenomena provided we accept this approximation. But Einstein was not interested in an approximately true relativity principle. He wanted an exact theory. That is why the relativity principle is taken as a postulate.

When I first learned about this all those years ago, this assumption seemed quite natural to me, as it surely did to Einstein. In a sense, at the time that Einstein proposed it, it was in the air. Whether he was

---

[3]They used the fact that a so-called parallel plate capacitor would, if suspended, change its orientation if it was in motion with respect to the aether. It would try to line itself up perpendicular to the motion. This would have been observable if a stationary aether existed. Interestingly this experiment was suggested by Fitzgerald—he of the contraction.

influenced by Poincaré, for example, or just his intuition, I do not know. On the other hand, the constancy principle seemed quite mad to me. Just think what it means. In our daily experience, if two objects approach each other their relative velocity is the sum of their individual velocities. That is why a head-on crash is so devastating. But this reasoning, the constancy principle asserts, does not apply to light. If a light beam approaches you, and you are moving towards it, its speed is the same as if you were at rest. If two light beams approach each other each with a speed c, their relative speed is still c. You can't catch up with a light ray. This seemed to me, as I said, quite mad. So I asked Professor Frank how Einstein could ever have thought up such a thing. There is no clue in his paper. I did not understand the answer Professor Frank gave me at the time. I did not then know enough physics. But I remembered his answer, and gradually came to understand it. He said it was in the mismatch of the physical dimensions of electric and magnetic fields. Here is what he meant. You recall from Chapter 1, that Lorentz derived from the Maxwell equations the law of force for a charged particle like an electron, moving in electric and magnetic fields. I noted that the force due to an electric field on, say, an electron is proportional to the strength of the electric field which I can call E. On the other hand, the force due to the magnetic field, H, is proportional to the strength of the field and the speed v at which the electron is moving. Because of the way these two terms enter the force equation, they must have the same physical dimensions. They must have the dimensions of a force. This is only possible if the dimension of the electric field $[E]$ equals the dimensions of some velocity $[v]$ times the dimensions of H, $[H]$. Thus $[E] = [v][H]$ or $[E]/[H] = [v]$. This velocity, whatever it turns out to be, has a universal significance. It is not measured with respect to any motions. It is an intrinsic parameter of the theory. But what is it?

This question was answered in 1857, by a beautiful experiment performed by the German physicists, Rudolf Kohlrausch and Wilhelm Weber. They made use of a device that is called a "Leyden jar"–because at least one of its mid 18th-century inventors was at the University of Leyden. In its original form it was just a glass jar partially filled with water. It was soon realized that if you simply coated the inner and outer surface of the glass with a metal coating that conducted electricity the same function could be fulfilled, but better. A metal wire is inserted through an opening in the top and attached to the inner wall. Then a negative charge is generated on the inner wall through the wire. This attracts positive charges from the outer metallic coating and the two charges, which cannot interact through the glass, are built up and maintained. A very hefty charge was stored for several hours in these early devices. What Kohlrausch and Weber did was to measure this charge in two ways—one involving electric fields alone, and the other involving magnetic fields. Basically, one can measure the size of the former by seeing how test charges are attracted or repelled by the charge on the jar. To measure the latter, they discharged the electric charge on the jar through a wire which produced a current. Associated with this current there is a magnetic field which can be measured by seeing how it affects compass needles and the like. Comparing the two measurements they found that the velocity that entered the field dimensions had the value of about 3.1×10^{10} cm/sec—the speed of light! I do not know how Professor Frank concluded that this was Einstein's reasoning. Maybe he asked him. Although I can't be sure, it seems plausible to me.

We are now in a position using these postulates to do some Einsteinian relativity. I will not follow the steps in Einstein's paper which require more mathematical reasoning than is appropriate here. Instead I will analyze the behavior of a particularly simple clock—the light clock. It is simple because in relativity the propagation of light follows a very simple law. It always propagates with the speed c. Figure 2.5 illustrates what I have in mind.

As I mentioned, the "clock" in question takes advantage of the fact that light in the vacuum obeys a very simple law of propagation. If the propagation is in the x-direction then it obeys the equation $x = ct$, where c is the speed of light in the vacuum. As we have said, Einstein's postulate is that every observer moving uniformly will measure the same speed of light, c. The clock consists of two mirrors placed a distance w from each other. If the mirrors are at rest with respect to each other and we have

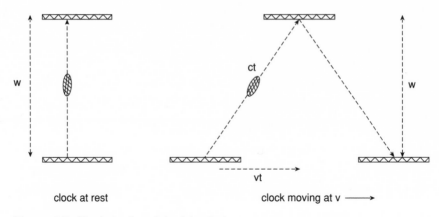

clock at rest   clock moving at v ⟶

Figure 2.5. The behavior of the light clock.

a light source at one of them, then the time it takes for a round trip is $2w/c$. This is the "period" of the clock–the basic time unit. Now suppose that the lower mirror is in motion to the right with a speed v. We could just as well assume that the lower mirror is at rest and the upper mirror is moving to the left. The analysis would be the same.

With the movement of the lower mirror an observer attached to the "rest system" will see the light follow a triangular path as shown in the second portion of the figure. This path is longer than the first path. If we call the time it takes to make one leg of the trip t as measured by this observer, then the period associated with this clock–the time it takes to make the round trip, again as measured by the resting observer is $2t$. It is clear from the drawing that this period is longer than $2w/c$ which is the period of the rest clock. So we know without doing any calculation that the observer at rest will argue that the moving observer's clock is running **slower!** In relativity this is called "time dilation" and it has profound consequences. To make this quantitative we use the Pythagorean theorem. In using it we will assume that nothing happens to the distance at right angles to the motion. This can be demonstrated with a relativity argument that I will not go into here. As mentioned Chapter 1, this point seemed to elude Lorentz. Thus one leg of the triangle is w. If $2t$ is the time dilated period, then the other leg of the triangle is vt, which is the distance along the x-axis the clock travels in time t. The light, however, travels along the hypotenuse. By the Pythagorean theorem this length is given by $\sqrt{w^2 + v^2t^2}$. But this must equal ct, the distance the light travels. Thus $ct = \sqrt{w^2 + v^2c^2}$ or solving for t we have $t = \frac{w}{c} \times \frac{1}{\sqrt{1 - v^2/c^2}}$. Let me call the rest period P_0, the time it takes the light to go back and forth in the rest system. Thus $P_0 = 2w/c$. If P is the period of the moving light-clock, as measured by the resting observer, then $P = \frac{P_0}{\sqrt{1 - v^2/c^2}}$, which reflects the extra distance the light goes along the hypotenuse. I keep insisting on putting in this reference to the resting observer, since the other observer will claim to be at rest and the logic is then reversed. The two situations are symmetric. Note that since $\sqrt{1 - v^2/c^2} \leq 1$ we have $P \geq P_0$. This is the "time dilation." How big an effect is this? Let suppose that v is nine-tenths the speed of light. This is a huge speed for everyday life, but not for particles in high energy accelerators. They go a lot faster than this. With $v/c = .9$, the period of the moving clock is somewhat more than doubled. Can we test this result? Yes, and I will give two examples.

The first involves the so-called elementary particles. Most of them are unstable. Eventually they decay into stable particles. Quantum mechanics, which tells us what we can know about these decays, cannot predict when a given particle–the one you have in your hand–will decay. It can predict the average lifetime of the particles. On the average they will decay at some time t, their so called "life-time." But whose time is it? Which frame of reference? If we measure the life–time with a light clock at rest with the particle we will get one answer, but if the particle is moving with respect to the laboratory then this time will be dilated. This means that the track of the particle that we observe is longer than it would have been be if there was no time dilation. It has more time to travel. This effect has been verified countless times in elementary particle laboratories. The second example involves an amusing irony. It deals with an aspect of the Doppler shift.

In Chapter 1, I described a device Mach invented to demonstrate the Doppler shift which, even several years after it had been postulated theoretically, was still regarded with some skepticism. Mach, it will be recalled, attached a tube, which functioned as a whistle, to a rotating wheel. If he stood in the plane of rotation he could hear the change in pitch as the wheel rotated. But, if he stood at right angles to the wheel, there was no shift. We would summarize this by saying that in classical physics there is no "transverse" Doppler shift, no shift at right angles to the motion. What about relativity? If a light wave goes by we can imagine measuring the time it takes for the passage of the wave say from crest to crest. This passage we can call the "period" of the wave. If we are at rest with respect to the source of the wave we can use the resting light-clock to calibrate this period, which I will call P_0. But what if the source is moving? We will then have to use the moving light clock with the longer path. So to an observer at rest in the laboratory, this period will be time dilated. The relativistic Doppler shift involves both this time dilation and the Lorentz contraction. Figure 2.6 depicts a series of waves emanating from a point source. As depicted, the source is in motion to the right. These waves in the rest frame have wave fronts that are spheres. If the source is set in motion then there will be, for the rest observer, a Doppler shift whose size and sign depends on the angle from which you observe the wave relative to the motion of the source. If you observe it head–on you will see a shift to higher frequencies and shorter wave lengths–a "blue shift." But if you observe it at right angles–the transverse Doppler shift–the period will be dilated by the factor $\frac{1}{\sqrt{1-v^2/c^2}}$. If we call this period P, then the Doppler

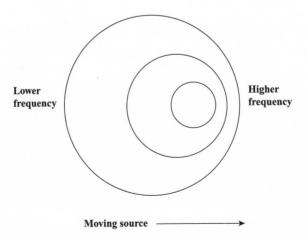

Figure 2.6. The Doppler shift illustrated using a series of waves.

shifted wave length is given by $\lambda = Pc = \dfrac{1}{\sqrt{1 - v^2/c^2}} P_0 c$. Since the square root is always less than one, the transverse Doppler shift lengthens the wave length–a "red shift." This is an effect that only shows up at order v^2/c^2, so it is absent in classical physics. Because v/c is, as a rule, tiny, it is a very tiny effect. But remarkably, it was measured by two Bell Telephone Laboratory physicists, Herbert Ives and G.R. Stilwell. who published their result in the Spring of 1938. Even more remarkably, they did not believe in the relativity theory–this was 1938!–and were hoping to disprove it. In fact they attempted alternate non-relativistic explanations long after the experiment was done.[4]

I would like to turn now to a discussion of the relativity of simultaneity–the Einsteinian version. This discussion will be qualitative. Later on, we can put in a few formulae. Recall the Newtonian discussion with the train and the lightning bolts. We will use the same set-up here. We said that the lightning bolts struck the embankment simultaneously at a time we can call $t = 0$. This means that from the point of view of the embankment light arrives at the mid-point at the same time. It is from this observation that we would argue that the striking of the lightning bolts represented simultaneous events. Even the aether theory people would admit that the observers at the midpoint on the train would not see the arrival of the lightning bolts at the same time. What they should have said is that this shows that simultaneity is relative, and that should have raised questions about the absolute nature of time. But it did not, with the exception of Poincaré, who seemed to understand the issues, although he did not invent a new mechanics. But as Einsteinian relativists, how should we view things?

---

[4]For those of you who know a little of mathematics the formula that gives the Doppler shift as a function of angle-angle measured by the observer is $P = P_0 \dfrac{1 - v/c \cos(\varphi)}{\sqrt{1 - v^2/c^2}}$. At ninety degrees this reduces to what we had before. Note that at zero degrees the formula reduces to $P = P_0 \sqrt{\dfrac{1 - v/c}{1 + v/c}}$ which is a blue shift. Stilwell and Ives tested the angle formula at zero and one hundred eighty degree using hydrogen atoms that moved with speeds the order of 10^8 centimeters per second and emitted light in all directions. They confirmed the general formula for these angles as well.

We have the same set-up with the lightning bolts striking at a time, say $t = 0$ as measured by the embankment clocks, at two points equidistant from the mid-point of the train. We now may ask, on what will the embankment observers and the train observers agree, and on what will they disagree? Both will agree that the lightning struck at points equidistant from the center of the train. However, we had better not assume that these distances are the same for both sets of observers. Remember length contraction. Both will agree that the speed of light is c. Both will agree that light from these strikes will not arrive at the midpoint of the train simultaneously. But the two sets of observers will have different explanations. The embankment observers will say that on one side the train was moving towards the light and on the other side away from it. Even though the velocity of light is the same in both directions, the distance traveled differs, so the light will arrive earlier at the train's midpoint on the side on which the train is moving toward the light. The train observers will have a different explanation. They will say that they are at rest, while the embankment is moving to the left. Our clocks, they will maintain, are all synchronized with each other, but are not synchronized with the embankment clocks. According to these embankment clocks, lightning struck simultaneously. But this is what *their* clocks say. Their clocks say that the lightning did *not* strike simultaneously. Our clock on the right reads an earlier time when the lightning bolt struck there than the clock on the left reads. That is why that light from the right got to the center sooner. It had a head start. Both sets of observers, embankment and train, are correct, and if they do the analysis consistently, will agree on the answer.

In this respect, consider the example of the decaying particle. If we call the life-time as measured on a system that is moving with the particle the "proper life-time"—t_0—then the time observed in the laboratory is $\frac{t_0}{\sqrt{1-v^2/c^2}}$ so that the distance traveled in the laboratory is $d = v\frac{t_0}{\sqrt{1-v^2/c^2}}$. But how do things look from the point of view of the moving particle? The distance moved will in this system, be vt_0. This will be equal to "d," but measured in the moving particles' length units. The only way these two answers can be consistent is if these units are contracted by $\sqrt{1 - v^2/c^2}$ or, in other words, $d\sqrt{1 - v^2/c^2} = vt_0$, which yields the same result.

Thus, in one system it is time dilation that produces the answer and in the other it is the Lorentz contraction.

This complementary action of space and time is typical of relativity. We can see it at work in another example. This one will again involve the train. Suppose we want to measure its length. If we are riding on it, there is no problem. We can take a tape measure and do the job. This length we will call the "proper length"–L_0. The problems arise when we ask how an observer on the embankment will measure this length. I will present two suggestions. You will then get the idea and can come up with your own methods. If you find one that produces a different answer you should book your hotel in Stockholm for the Nobel Prize awards. One way that looks pretty simple is to station an observer on the embankment with a clock. When the front of the train passes, start the clock. When the back passes, stop it. Then take the time interval, multiply by the speed of the train and you will have its length. Or will you? By now you will see the problem–time dilation. The two observers will disagree on the time interval with the result that the length expressed in the units of the moving train will be contracted. Here is another method. Suppose that at a certain time, say $t = 0$ as measured on the train, someone at the front and back end of it agree to simultaneously put down a marker on the track. At their leisure, observers on the embankment can measure the length between these markers. The problem is that the embankment observers will insist that, in their time units, the placement of these markers by the train people was not done simultaneously. The second placement came after the first. Because of this, the length of the moving train as given by this method is again contracted. Space and time play complementary roles.

A beautiful example of the interweaving of space and time in relativity is the relativistic treatment of the phenomenon of stellar aberration. You will recall that if a telescope on the moving Earth is used to observe a star, it must be tilted slightly because the light rays from the star appear to arrive at an angle. If the star is treated as a point source of light then the light that is emitted emerges as spherical waves. The surfaces of the advancing light rays are spheres. But the stars are so far away the by the time the light reaches us these spheres are huge. For all practical

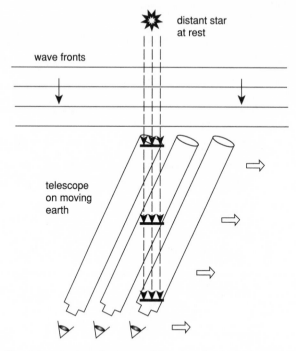

Figure 2.7. The interweaving of space and time in relativity is shown in the relativistic treatment of stellar aberration.

purposes we see so little of the surface that it appears flat. So we can regard these waves as plane wave surfaces as shown in Figure 2.7.

In the aether theory, a moving star–as opposed to a moving earth–would not show aberration (see Figure 2.8). This would be a violation of the relativity principle which demands that the two situations be symmetric. This does not surprise us because we know the aether theory violates the relativity principle. In Einstein's theory what happens is that the various points on the planar surface do not arrive at the telescope simultaneously when the star is in motion (see Figure 2.9). Hence there is aberration and the two situations, star at rest, Earth moving, and star moving and Earth at rest are perfectly symmetric as relativity demands.

To make relativistic ideas quantitative we must, as Einstein did, find the transformations of space and time that generalize the "Galilean transformations"–Professor Frank's term—$x' = x - vt$, $t' = t$. I will always deal with motions in the x direction so that the other two

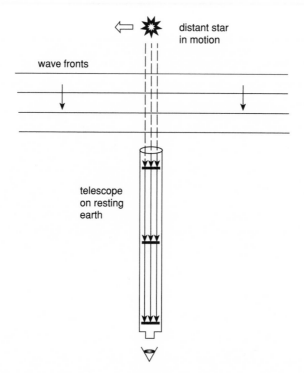

Figure 2.8. As Figure 2.8 shows, no aberration occurs in the aether theory.

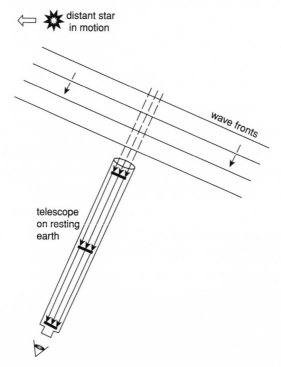

Figure 2.9. In Einstein's theory of relativity, aberration does occur.

coordinates, y and z are not affected. As I have mentioned, the first person to write down such a generalization, but with a different motive, was Lorentz. His motive was to simplify calculations of the electrodynamics of moving bodies by transferring the problem to a coordinate system in which the body was at rest. He referred to this as finding "corresponding states." These transformations could not be the Galilean transformations since those transformations do not preserve the form of the Maxwell equations when you go from one system to the other. He found, as I noted in the previous chapter that, if he ignored higher powers of v/c, a transformation that worked was $x' = x - vt$ and $t' = t - v/c^2x$, with a new time, t', which he called the "local time" because it involved the coordinate x. He did not attach any physical significance to this time. It was just a mathematical artifact. As I also mentioned, in 1904, Lorentz published his final prerelativity paper on this. In it he wrote down the transformation for the general case where v/c is not small. But, in fact, it is too general. He has a scale factor that allows distortions in all directions. Since he does not adopt the relativity principle he has no argument for setting this factor equal to one, so he is stuck with the distortion.[5] But, in recognition for his pioneering work, the transformations I am about to write down, without the scale factor, are called the "Lorentz transformations" by everyone. Such is history. Incidentally, if you set the speed of light equal to infinity in the Lorentz transformations you revert to the Galilean transformations of Newtonian physics. This is not surprising, since if electromagnetic communication was instantaneous, there would be no relativity of simultaneity. You would have the absolute

---

[5] Einstein disposes of this factor in his 1905 paper. He argues that if you make a Lorentz transformation, followed by its inverse–that is, you transform to a system moving with v, followed by one moving with $-v$–you get back to the original rest system. Allowing for the fact that l, the distortion factor, might be a function of v you get the condition that $l(v)l(-v) = 1$. If you assume from symmetry that $l(v) = l(-v)$ you get $l^2 = 1$. The negative root is ruled out by continuity so you get the answer. Why Lorentz did not invoke this argument I do not know. Poincaré, had his own proof.

time of Newtonian physics and hence the Galilean transformations of space and time.

Since Einstein's 1905 paper, a veritable cottage industry has grown up devoted to deriving these transformations. Rigor, or the lack thereof, is the hallmark of these various derivations. I am not going to present a real derivation. That would take us too deeply into mathematical waters. You can find real derivations galore in innumerable places. Many of them follow Einstein's original method, as expressed in his 1905 paper. It uses the notion of spherical light waves emanating from a source. These wave fronts propagate as a spherical surface. What Einstein noted was that, according to the relativity and constancy principle, if there is a moving observer at the source than he, or she, will also report a spherical light wave expressed in the moving system's space-time coordinates. The relativity transformations must reflect this. Indeed, making this demand and adjoining some other plausible mathematical assumptions, was how Einstein derived the Lorentz transformations. The transformed time, t', which I will shortly write down, is not in Einsteinian relativity a mathematical artifact as Lorentz maintained. It is something measured by clocks. Furthermore, Einstein's derivation has nothing to do with any model of matter, such as the one that Lorentz favored. Einstein's derivation is based on very general assumptions about space and time. All acceptable models of matter must conform to these assumptions.

As I said, I am not going to present a derivation of the transformations. Rather it will be a "derivation." It is the one I first learned from Professor Frank. It assumes the Lorentz contraction. That is why it is a "derivation" and not a derivation. The real derivations do not make this assumption and indeed the Lorentz contraction emerges as a consequence. The "derivation" is based on the mismatch of units in the equation $x' = x - vt$. We have learned that primed units are Lorentz contracted. To make the units match we should write $\sqrt{1 - v^2/c^2}x' = x - vt$. Thus $x' = \frac{x - vt}{\sqrt{1 - v^2/c^2}}$. This is the first of the Lorentz transformations. It replaces the Galilean transformation $x' = x - vt$ to which it reduces when the speed of light is infinite. We will now turn to the time transformation.

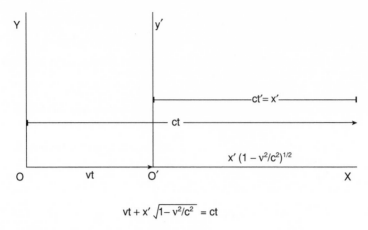

Figure 2.10. The Lorentz time transformation.

We imagine a situation in which a light beam is generated at a time $t = t' = 0$ at the common origins of the two systems shown in Figure 2.10. We suppose the light beam moves towards the right. After a time t in the unprimed system it arrives at the point x. It has thus moved a distance, as viewed in the unprimed system, which is given by $x = ct$. If we translate this into the prime language the same distance is given by $vt + x'\sqrt{1 - v^2/c^2}$ or in other words $ct = vt + x'\sqrt{1 - v^2/c^2}$. But $x' = ct'$, light propagates from the origin of the prime system with the speed c. Putting this in and solving we find that $t' = \dfrac{t - v/c^2 x}{\sqrt{1 - v^2/c^2}}$. To the leading order in v/c, this formula is identical to Lorentz's 1895 version of the "local time." If we let c become infinite here we have $t' = t$, absolute Newtonian time, If we complete the list with the transformations $y' = y$ and $z' = z$, we have the full set of Lorentz transformations—at least the Einsteinian version.[6]

Before I discuss the rest of Einstein's paper, here are a few applications of the Lorentz transformations. First there is the relativity of

---

[6]A courageous reader can verify using the Lorentz transformations that $x'^2 + y'^2 + z'^2 - c^2\tau'^2 = x^2 + y^2 + z^2 - c^2\tau^2$. If we set either side equal to zero we have the equation for a propagating spherical light wave. Most derivations, including Einstein's, start with this invariance and using simplifying mathematical assumptions arrive at the Lorentz transformations. With our "derivation" we have gone about it backward.

simultaneity. Let us go back to the train and the lightning bolts. In the embankment system the bolts strike at points, say, plus and minus L at $t = 0$. This means that the two places, $+L$ and $-L$ are equidistant from the origin of coordinates and the lightning bolts strike these points simultaneously. The Lorentz transformations tells us that, in the prime coordinates, the coordinates of the moving system, the lightning strikes at points plus and minus $\dfrac{L}{\sqrt{1-v^2/c^2}}$, which are also equidistant from the origin. If we apply the time transformation then left-hand lightning bolt strikes at a time $v/c^2 \times \dfrac{L}{\sqrt{1-v^2/c^2}}$ while the right-hand bolt strikes at $-v/c^2 \times \dfrac{L}{\sqrt{1-v^2/c^2}}$. Thus, in this frame, the two bolts do not strike simultaneously. The right-hand one strikes first — this is what the negative sign means — which corresponds to our intuitive idea that the light from the bolt towards which we are moving gets to the center first. It gets a head start.

The second example is taken directly from Einstein's paper. It involves the time dilation. We suppose a clock at the position x is moving to the right with a speed v. Thus the equation for the clock is $x = vt$. If we put this into the time Lorentz transformation we have $t' = t\dfrac{1-v^2/c^2}{\sqrt{1-v^2/c^2}} = t\sqrt{1-v^2/c^2} = t - \left(1 - \sqrt{1-v^2/c^2}\right)t$. In the last term we have added and subtracted t to make the time delay more transparent. If v/c is small then this expression is approximately $t(1 - 1/2v^2/c^2)$. In other words, the time delay is about $1/2v^2/c^2$. Einstein suggested a fanciful experiment to test this. You put one clock at the North Pole and another identical one at the Equator. At the Equator the Earth is going around at a speed of about 0.46 kilometers per second. So the time dilation effect is predicted to cause a time delay of about one part in 10^{12}. What is amusing about this, is that if Einstein had been able to carry out this experiment, he would have discovered that there was **no** time delay! He did not realize this at the time, but there is an effect of gravity on the clocks that just compensates for this time dilation. In fact, the experiment has been done with atomic clocks flown in airplanes and this result is confirmed.

Next I want to turn to the velocity addition theorem in relativity, the one that replaces the Newtonian one that was based on absolute time. To find it, we consider our usual diagram of the two coordinate systems. The prime system moves to right with the speed v. We will assume that the point x moves to the right with the speed u; $x = ut$. Expressed in terms of prime coordinates we have $ut = vt + x'\sqrt{1 - v^2/c^2}$. In the prime system x' obeys the equation $x' = w\,t'$, where w is the speed of x' in the prime system. This is just the speed we are looking for. In the Newtonian case it is simply $u - v$. If we solve for w here, we find, $w = t/t'\frac{u - v}{\sqrt{1 - v^2/c^2}}$. The next step is to find t/t'. From the time Lorentz transformation we have with the same substitution for x; i.e., $x = ut$, $t/t' = \frac{\sqrt{1 - v^2/c^2}}{1 - uv/c^2}$. This gives us our addition formula, i.e., $w = \frac{u - v}{1 - uv/c^2}$. This wonderful formula has an interesting property. Suppose you let $u = c$. Then you have $w = c$. The speed of light looks the same to the moving observer. You cannot catch up to a light beam. You can rewrite the formula for two speeding objects that approach each other by changing the sign of v. If you then have two light beams that approach each other, their relative speed is still c. You can't go faster than the speed of light.[7]

In this respect I want to tell you a limerick. It is not the funniest limerick I know, but it is the funniest limerick about relativity that I know. It goes,

There was a young lady named Bright,

Who could travel faster than light.

She started one day,

---

[7] I am oversimplifying here. It is better to say that no information can be transmitted faster than the speed of light in the vacuum. There are subtle issues, of which I will not go into in detail here, as to what the speed of a wave means. You can think of a single wave that is represented by a simple trigonometric function. You might imagine that the whole wave is being displaced. On the other hand it might be that successive portions simply move up and down, something like the wave in a cheering section in a football stadium. In the latter case nothing is being displaced and the speed can be anything. To transmit information you must be able to modulate waves. This involves introducing packets of waves. These packets move with what is called the "group velocity." There are exotic media in which this group velocity can exceed c. But for these media it can be shown that no signal-transmitting information can travel faster than c. In the early years of relativity the nature of wave propagation received a good deal of attention. It has again become an active branch of research.

In a relative way,

And arrived the preceding night.

Neither Miss Bright, nor anyone else, can travel faster than light. But there is an interesting point here. Can you, by making Lorentz transformations, change the time order of events? If one event follows another in one frame of reference, is there another frame of reference reached by Lorentz transformations that reverse the order? If so, in one frame you might see a bank robbery followed by an arrest, while, in another, it would be an arrest followed by a bank robbery. In relativity this sort of cause and effect reversal is impossible. If you stick to speeds less than light you cannot do this switch—fortunately. Causality is preserved. As a friend of mine says about rock climbing, relativity may be ridiculous, but it is not absurd.[8]

In the first part of Einstein's paper, which he calls the "Kinematical Part," he discusses what we have been discussing. This is followed by the "Electrodynamical Part." Here Einstein makes use of the Maxwell equations. He first shows that these retain their form under Lorentz transformations, which is to be expected because this was one of the motivations for deriving them in the first place. The notation that Einstein uses in this discussion is so cumbersome that strong men have been known to weep when they read this part of the paper. The logic and mathematics are, needless to say, correct. He then deals with such things as the Doppler shift. The tenth section of this part needs our attention. It is called "Dynamics of the Slowly Accelerated Electron." The Special Theory of Relativity, which has been our subject, is "special" because it deals only with transformations among systems that are moving uniformly with respect to each other—so-called "inertial systems." But to make it a realistic physical theory we must allow forces to act. But forces produce accelerations, so how to proceed? The idea is this. An acceleration is a change in velocity. But at any instant of time the electron, for

---

[8]There has been some theoretical speculation as to whether particles that always move faster than light—so-called "tachyons"—can exist. So far there is no evidence for such exotic objects.

example, will have some velocity. In the next instant of time it will have a different velocity if it is being acted on by a force. What we do is, at an instant of time, we make a Lorentz transformation from, say the laboratory, to a system moving with that instantaneous velocity. In this system we apply Newton's law, or at least the relativistic version, and then transform back to the laboratory to find the equation of motion. In his paper, Einstein's treatment of this is somewhat peculiar. So I will not try to reproduce it. Rather I will sketch how we would do this now. Soon after his paper was written, Einstein accepted this improved formulation. The first thing that we would do is to show that in relativity the momentum, p, is not mv, where m is the mass. The correct expression in relativity is $p = \frac{mv}{\sqrt{1 - v^2/c^2}}$. The 'm' that appears here is the so-called "rest mass." This is the mass an object has if brought to rest. There are two interesting things to note about the expression, $\frac{m}{\sqrt{1 - v^2/c^2}}$. In the first place, as v approaches c, it gets larger and larger and finally becomes infinite when $v = c$. When I first learned about this, those many years ago, I was very puzzled. I had learned the Newtonian definition of mass as the quantity of matter. How could this possibly increase? Were new atoms somehow added? But this is not at all what mass means in this context. Mass is a measure of the difficulty that a given force has to accelerate a particle–a measure of its "inertia." What this increase in mass means, is that, as a particle approaches the speed of light, it becomes more and more difficult for it to be accelerated. You can never accelerate it to the speed of light. This is as it should be, because no massive particle can move with the speed of light. This brings us to the second point. We know that there is at least one thing that moves with the speed of light–namely light. In the final chapter of this book I will point out that, in many contexts, light behaves as if it is made up of particles–"photons" we call them now. At the speed of light the denominator in the momentum expression is zero, so the momentum makes no sense for such a particle, unless the numerator is also zero. This can only happen if the mass, m, is zero. Thus the photon must be a **massless** particle that moves with the speed of light. You can never bring it to rest and you can never catch up with it. Since the relativistic momentum depends on the velocity in this more complex way, Newton's law in the form $F = ma$ does not hold in relativity. The form you have to use is $F = \frac{\Delta p}{\Delta t}$. This makes the equations a lot harder to solve. When Einstein applied this new dynamics to electrons he noted that the force that acted on them depended on the state of motion of the electrons. An electron in a rest system might be subject only to an electrostatic force—Coulomb's law. But if you observed the same electron from a moving system it would be subject to a magnetic force as well, as a consequence of the Lorentz transformations. The two forces are part of a common whole—"electromagnetism." It is the first example of what has become the Holy Grail of physicists—the unification of all the forces.

In the beginning of his paper, Einstein cites a dilemma. On several occasions he said that it was one of the things that impelled him towards relativity. It is along the lines of what I have just been discussing. Figure 2.11 has a permanent magnet and a conducting loop. The loop is made of some kind of wire that conducts electrons. We can now, following Einstein, consider two situations. We can imagine moving the magnet with a constant speed v in a direction out of the diagram. The laws of electricity and magnetism that Einstein was familiar with from

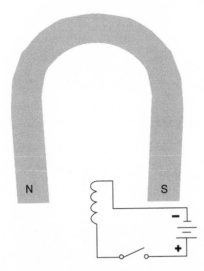

Figure 2.11. Einstein's dilemma illustrated with a permanent magnet and a conducting loop.

his student days, implied that an electric force would be induced by the moving magnet. This force acts on the electrons in the loop, producing a current in the loop. Now we can consider a second experiment in which we move the loop with the same speed in the opposite direction. The Lorentz law of force tells us that there is a magnetic interaction on the electrons which again sets up a current in the loop. This current is identical to the one set up in the previous case. Indeed, the current depends on the relative velocity of the magnet and the loop, which is the same in either case. What troubled Einstein, he tells us, is that the same phenomenon appeared to have different explanations—electric or magnetic—depending on which part of the apparatus is being moved. He found this intolerable. A modern physicist examining this situation would be at a loss to understand what the fuss was about. But that is because we now accept the theory of relativity. As I have indicated, this implies that electricity and magnetism are the same phenomenon—electromagnetism—and how these forces manifest themselves depends on which coordinate system we are in. This is what Einstein taught us.

You will not find the most famous equation in relativity in Einstein's first paper. This is, of course, $E = mc^2$. It is there implicitly, but no attention is called to it. But Einstein wrote a second small paper with the ungainly title, "Does the Inertia of a Body Depend on its Energy Content?" He considers a particular example. He imagines a "body" that can emit a pair of light waves in opposite directions. He examines this process in two systems of coordinates moving relative to each other. By comparing the results he concludes that if the light has carried off an energy L—his notation—then the "body" has suffered a mass loss of L/c^2. We know that c is a huge number. So a modest mass loss can be equivalent to a large amount of energy. The success of nuclear weapons is an unfortunate testimony to this. The first thing that struck me when I heard about $E = mc^2$, was how did such a large amount of energy manage to remain concealed for such a long time. Why had nobody noticed it? The explanation for this is not hard to come by. Until the beginning of the twentieth century, typical processes one considered were billiard ball-like collisions in which the same particles left the collision as entered it. The masses were the same initially and finally and the mass-energy simply did not affect the energy balance. By the time Einstein wrote this paper, radioactive elements were being studied. One of their great mysteries was where did the energy they were giving off come from? They seemed to give off an unlimited amount. At the end of his paper Einstein writes, "It is not impossible that with bodies whose energy-content is variable to a high degree (e.g., with radium salts) the theory may be successfully put to the test." It was, and it passed. A radioactive decay involves a loss of mass of the decaying particle. Put more precisely, the particles into which it decays are less massive and this mass difference is available as energy.

I now want to turn to how the relativity theory was received and then finally to a brief summary of what happened to the physics after 1905.

✎ REACTIONS

I do not believe that there is any man now living who can assert with truth that he can conceive of time which is a function of velocity or is willing to go to the stake for the conviction that his "now" is another man's future or still another man's past.
— W.F. Magie, Princeton University physics professor, in his Presidential Address to the American Association for the Advancement of Science, 1911

In the summer of 1949, Professor Frank gave me a job. He was preparing an article on how Einstein's relativity theory had been received by both scientists and philosophers. He had a small grant and was able to pay me a bit for doing some library research. One of the sources I read was the French philosopher Henri Bergson who, for reasons beyond my comprehension, was awarded the Nobel Prize in Literature in 1927. While I did not understand that much about relativity, I was persuaded that I understood it a lot better than Bergson. Here is a tiny fraction of what he wrote in "Time and Free Will"[9] an essay he published in 1889, but whose tenets he maintained long after relativity had been discovered.

> What duration is there existing outside us? The present only, or, if we prefer the expression, simultaneity. No doubt external things change, but their moments do not *succeed* one another, if we retain the ordinary meaning of the word, except for a consciousness which keeps them in mind. We observe outside us at a given moment a whole system of simultaneous positions; of the simultaneities which have preceded them nothing remains. To put duration in space is really to contradict oneself and place succession within simultaneity. Hence we must not say that external things *endure*, but rather that there is in them some inexpressible reason in virtue of which we cannot examine them at successive moments of their own duration without observing that they have changed. But this change does not involve succession unless the world is taken in a new meaning on this point we have noted the agreement of science and common sense.

I duly noted quotations like this—the meanings of which, if any, were beyond me—for Professor Frank. I have no idea what he did with them. In 1921, Bergson published his *Durée et Simultanéité: A Propos de la Théorie d'Einstein*. Einstein had the patience to point out the mistakes in Bergson's book, after which Bergson had the good sense to retire from this particular field.

Einstein sent his 1905 relativity paper to the *Annalen der Physik* in Berlin. The editor at the time was Paul Drude—a physicist known for his work on classical optics. But there was an editorial council on which the very distinguished German theoretical physicist, Max Planck, served. It was Planck who was given the paper to referee. We will learn more about him in the last chapter, but let me note here, that he was a deeply conservative man, both in his science and everything else. One might

---

[9] *Time and Free Will: An Essay on the Immediate Data of Consciousness*, trans. by F.L. Pogson. London: George Allen and Unwin, 1910.

imagine him rejecting Einstein's paper, because it was so radically new. There was no outside refereeing system, so that this would have been the first and last judgment. Instead, Planck embraced it with enthusiasm. Fortunately, he did not ask for changes, such as putting in references and explaining previous work on the aether. This would have helped the historian, but dated the paper, which is as fresh as when Einstein wrote it. Planck went further. His did his own work on the theory right away and was probably the first person, beside Einstein, to publish a relativity paper. Planck had a new assistant named Max von Laue, who had taken his degree with him in 1903. When Laue took up his post in 1905, one of the first talks he heard was Planck's colloquium on Einstein's relativity theory. Either on his own, or with the encouragement of Planck, Laue decided to visit Einstein in Bern, in 1906. He was astonished to find that they were the same age. Laue seems to have been the first contemporary physicist Einstein had ever met. Laue too began working on relativity, and published the first technical monograph on the theory in 1909. In 1914, he won the Nobel Prize in physics for his work on x-ray studies of the structure of crystals. When the Nazis came to power, Laue was the only prominent German physicist I know of, who remained in the country, and was outspoken in his opposition to the regime. He insisted on teaching relativity as Einstein's discovery even though it had been categorized as "Jewish physics"–only to be taught if Einstein was not mentioned. After the war, a colleague of Einstein's in the United States was going to visit Germany. He asked Einstein if he would like to send his greetings to any of the German scientists. In reply, Einstein asked that his greetings be sent to Laue. The colleague asked if there were any other German scientists Einstein wanted to greet. He gave the same answer.

Apropos of Laue, I want to return briefly to something I discussed in Chapter 1–the Fresnel "dragging coefficient." It will be recalled that Fresnel produced what seems to have been an empirical formula that described the speed of light passing through a moving medium like water. It was expressed in terms of the index of refraction, n, of the medium which in simple cases is just the ratio of the speed of light in the vacuum, c, to the speed in the medium at rest, v, i.e., $n = c/v$. His formula claimed that the observed speed in a medium moving with a speed, w, would be given by $c' = v + (1 - 1/n^2)w = v + (1 - v^2/c^2)w = v + w - v^2w/c^2$. The reason why I have written the formula this somewhat odd way will be evident shortly. In the 19th century there were heroic efforts to derive this formula. Lorentz, for example, produced a complex derivation using the aether theory and his electrodynamical model of matter. While Einstein never did give a full accounting of what led him to relativity, in those accounts he did give, he mentioned the Fresnel formula. He never said why it was so important to him. Perhaps it was another bit of evidence of the uselessness of the aether concept. What is odd is that he did not discuss it at all in his 1905 paper. He seemed not to have realized that there was a one-line derivation using his velocity addition law. This was found by Laue in 1907. It goes like this. We want to add v and w relativistically. Thus $c' = \frac{v+w}{1+\frac{vw}{c^2}}$. But w is much smaller than c, so we have approximately $c' \approx v + w - \frac{vw}{c^2}v$ which is the same as the Fresnel expression. End of story. It has nothing to do with the aether, or models of matter. It is a general feature of the relativistic view of space and time.

The first experimental work on relativity was performed by the German physicist Walter Kaufmann who was at Göttingen, which was the mathematical capital of the world. He had been studying the behavior of fast electrons in electric and magnetic fields for many years. Such experiments enable one to measure the mass of the electron as a function of its velocity. By 1905, in addition to Einstein's theory, there were at least two rival models that predicted that there would be such an effect but, with a different functional form. Recall that relativity predicted that the mass would increase as $\frac{1}{\sqrt{1 - v^2/c^2}}$. In November of 1905, Kaufmann published his results. Of the three models he tested, what he called the "Einstein-Lorentz" model, came out the worst. There are philosophers of science[10] who assure us that what distinguishes science from other disciplines is that its predictions are falsifiable. On this basis, the relativity theory was dead. Indeed, this is what Lorentz thought and was in some state of desperation as to what to do next. He wrote to Poincaré that he

---

[10]Notably, the late Karl Popper.

was "at the end of his Latin"–meaning that he was stuck. And Einstein? He simply did not care about Kaufmann's experiment. He was sure that sooner or later it would go away. He did not make any public comment until he wrote a review article in 1907. Of the other theories that seemed to agree with experiment he noted, "However, the probability that their theories are correct is rather small, in my opinion, because their basic assumptions concerning the dimensions of the moving electron are not suggested by theoretical systems that encompass larger complexes of phenomena." Einstein was right. A few years later, better experiments confirmed his theory.

The next significant advance in relativity theory was made not by Einstein, but by his old mathematics teacher at the "Poly," Hermann Minkowski. Minkowski was born in what was then Russia in 1864, however he was educated in Germany. He taught in a few places, including Zurich, before he accepted a chair in Göttingen in 1902. Among his colleagues was David Hilbert, whom many people think was the greatest mathematician of the twentieth century. Hilbert also had a deep interest in physics and created some of the mathematical tools we still rely on. He organized a seminar on electron theory at Göttingen in 1905, in which Minkowski participated. Not long often he learned about relativity. Minkowski soon found a way of rewriting the theory. He presented it in a lecture entitled "Space and Time," that he gave in September of 1908, to the 80th Assembly of German Natural Scientists and Physicians in Cologne. It is one of the most remarkable physics lectures ever given by anyone. It transformed how we think about relativity. It put relativity on the map.

Minkowski's basic idea was that relativity is something that takes place in a four-dimensional space. Each "event" is characterized by three space coordinates x, y, and z and one time coordinate t. This was also true of Newtonian mechanics, but the time coordinate was not interesting. It has the same value in all the Galilean reference systems — absolute time. On the contrary, in relativity, the time variable–the fourth dimension–is interesting. Under Lorentz transformations it gets mixed up with space. Minkowski began his lecture by saying,

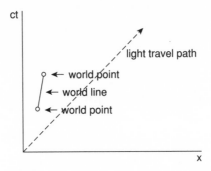

Figure 2.12. A space-time picture with x and ct axes at right angles.

The views of space and time which I wish to lay before you have sprung from the soil of experimental physics, and therein lies their strength. They are radical. Henceforth space by itself, and time by itself, are doomed to fade away into mere shadows, and only a kind of union of the two will preserve independent reality.

To picture what is happening, Minkowski introduced diagrams for what he called the "absolute world." You cannot plot four dimensions all at once, so Minkowski used one, or two, space dimensions and one time dimension. We will stick with one space dimension which simplifies things without losing anything essential. Instead of time it is customary to plot "ct" where c is the speed of light. This gives space and time the same physical dimensions and makes things more transparent to plot. Figure 2.12 is a space-time picture. You will notice the x and ct axes which are at right angles.

Because of our choice of units, light, which obeys the equation $x = ct$, moves along the 45° line. All trajectories—"world lines"–that represent motion at less than the speed of light move in trajectories at larger angles. No motion is represented by a straight line parallel to the time, or ct, axis. You do not get anywhere, but just get older. From the origin in Minkowski space you can, in two dimensions, draw what are known as "light cones." If you are at the origin your entire future is in the "forward light cone" that goes up the axis, while your past is in the "backward light cone." Outside the light cones you have the "absolute elsewhere" with which you can never communicate. See Figure 2.13.

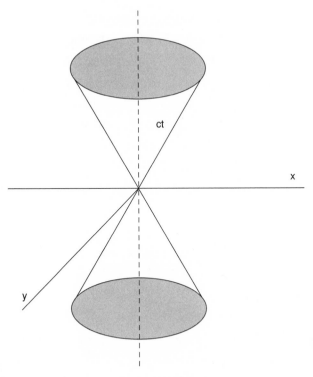

Figure 2.13. Light cones in two-dimensional space.

It is somewhat more difficult to see how Lorentz transformations are represented in the Minkowski diagrams. In essence what happens is that the two axes are tilted as shown in Figure 2.14. The essential point is that in the new coordinate system light will propagate with the equation $x' = ct'$, so the light signal will travel along a line that bisects the new time axis and the new space axis. Instead of axes we have drawn

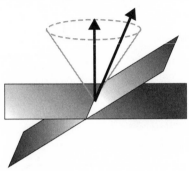

Figure 2.14. Planes of simultaneity in the Lorentz frames.

planes. Events located on any given plane will be simultaneous with all the other events on the plane. The relativity of simultaneity is reflected by the fact that in the two systems the planes of simultaneity are tilted with respect to each other. By working out the geometry you can reproduce all the algebraic results of the Lorentz transformations. A few months after giving this lecture Minkowski died suddenly because of a ruptured appendix. He was only forty-four years old.

Finally, I would like to mention the "traveling twins." No discussion of relativity is complete without the twins. Apparently the first person to state this "paradox" was the French physicist Paul Langevin in 1911. The first really satisfactory resolution was von Laue's a year later. Since then, forests have been cut down to supply the paper that has gone into discussing this. In essence it is the following. You have twins. One stays home at rest and the other goes on a round trip (see Figure 2.15). The human heart is a kind of clock. so we have time dilation. No one seems to have any trouble with the fact that, as seen by the resting twin, the other twin will have aged less on its return. This, everyone agrees, is a straightforward application of time dilation. The fun begins when one attempts to get the same result as viewed by the traveling twin. The two twins must agree on the aging of the traveling twin when they are re-united. Naively from special relativity we might at first think that the two situations are symmetrical so that the traveling twin will also report that the stationary twin has aged less which leads to an absurdity. But, in

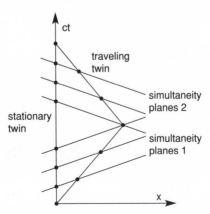

Figure 2.15. The Minkowski diagram of the travelling twin.

fact, their situations are not symmetrical. The traveling twin experiences acceleration. I will spare you a description of these learned arguments as to how this works out in detail. It is easy to wander off the deep end here. I think that if you keep your wits about you at the end of the day you will find that there is no paradox. The twins will agree on the aging. Professor Frank, when describing this, used to say, "Travel and stay young."

While I am on the subject of special relativity I would like to call your attention to some work that was done in 1959, by the Los Alamos physicist James Terrell. Before Terrell did his work, there were charming drawings of the Lorentz contraction showing people squished as they moved close to the speed of light. The implication was that if you could actually observe these people this is what you would see. By examining the relativistic optics, Terrell showed that this is not at all what you would see. In fact, if you looked at a meter stick in relativistic motion you would see an un-squished meter stick that was rotated. No one had thought to do this bit of physics before. Einstein was no longer alive, but one can imagine that he would have been very amused.

At first sight you might think that gravitation, which was the only other force being considered at this time, would be straightforward to include in the mix. You might approach it as follows. The force of gravitation is like the electrostatic force—Coulomb's law—in that it falls of with the square of the distance between, say, two gravitating objects. You might then do what you do in the Coulomb case, and Lorentz transform to a moving system. The problem is that you do not have anything like gravitational magnetism. The velocity dependent forces you generate this way do not correspond to anything observed. So you must use a totally different tack. Einstein described how, in November of 1907, he found the clue which over the next eight years led to the answer. At first sight what he says seems so bizarre that one is tempted to think that it may be one of his "krecks." Here is what he later said in a lecture: "I was sitting in a chair in the patent office at Bern when all of a sudden a thought occurred to me. 'If a person falls freely he will not feel his own weight.' I was startled. This simple thought made a deep impression on me. It impelled me toward a theory of gravitation."

Before I explain how this "simple thought" impelled Einstein towards a theory of gravitation, let me explain the thought. Imagine that you are standing on a scale on a platform weighing yourself. The platform gives way and you, and the scale, and everything else begin a free fall in the field of gravity. Every object, whatever its mass, will fall with the same acceleration. (I ignore effects of air resistance.) We demonstrate this in freshman physics by dropping a penny and a feather in an evacuated tube and watching them fall. This acceleration, g, is the famous 32 feet per second/per second. How do we derive this? We use Newton's $F = ma$. We won't worry about relativistic effects here. In this equation, F is the force of gravitation at the surface of the Earth whose radius we will call 'R.' Thus $F = \frac{GmM}{R^2} = mg$. Here "M" is the mass of the earth and G is Newton's gravitational constant that measures the strength of the gravitational force and g the acceleration If we cancel the 'm' from both sides of the equation we have $g = \frac{R^2}{GM}$. The mass of the object being accelerated has dropped out. Einstein noted that in making this cancellation we have implicitly assumed something about the masses. On the "mg" side of the equation, the mass is being used as a measure of the inertia. Einstein called this the "intertial mass," m_i. On the force side of the equation, the mass is a measure of the strength of the gravitational attraction. He called this mass the "gravitational mass," m_g. What has been assumed is that $m_i = m_g$. Einstein did not know at the time that a Hungarian Baron named Roland Eötvös, had made for decades, a career out of showing experimentally that these two masses were the same. The best answer he got was that they were, to one part in a billion. Now it is known that they are, to one part in a trillion.

Einstein realized that this equality of masses led to a new and different kind of relativity principle. This one involving accelerations. He later illustrated it with a little thought experiment. He imagined an observer inside a "chest"–a "spacious room"– somewhere in space. He cautioned that the observer should fasten himself to the floor with strings because otherwise he was going to fly up and hit his head on the ceiling when the chest was jarred. He then imagined a "being" who can pull up on the chest with a rope. The being accelerates the chest upward with the gravitational acceleration g. The occupants of the chest–now usually called the "Einstein elevator"–will find that they feel the acceleration. In fact it will be the same effect as if there was no "being" but the chest was in a uniform gravitational field. Things will appear to fall with the acceleration, g, and a scale will give your weight. This observation Einstein elevated to a principle–the Principle of Equivalence–the equivalence of a uniformly accelerated system and a uniform gravitational field. Being Einstein, he saw how to use this principle to draw some remarkable conclusions. Figure 2.16 gives an example. A laser beam is injected into our elevator– or rocket ship. The being pulls the elevator up with an acceleration, g. As the figure shows, when the beam exits the elevator it will be at a lower place. But we can view the same effect in a uniform gravitational field. Now what has happened is that gravity has bent the laser beam. When Einstein published his paper on this in 1911, "On the Influence of Gravitation on the Propagation of Light," he made a prediction. He said that the Sun would bend starlight and that this might be observed in an eclipse of the Sun because then you could observe starlight passing close to the Sun, which you could not do in broad daylight. He gave a figure for how much bending there would be–a tiny angle. In 1914, a group of German astronomers went to Russia to observe the Sun during an eclipse. The war broke out and they were lucky to get back to Germany, minus their equipment. They never did get to do the observation. If they had, they would have found an effect *twice* what Einstein predicted in his 1911 paper. Einstein had, by this time moved well beyond it.

There is a second consequence of the Principle of Equivalence I would like to tell you about before I describe Einstein's final theory of gravitation. We can imagine that we have an atom that emits light in our elevator. The "being" gets to work and the floor begins to accelerate towards the atom. This will produce a Doppler shift in the light–a blue shift as I have described it. But the Principle of Equivalence implies that the same effect can be produced if you put the atom in a gravitational field. This was first detected directly in an experiment done at Harvard by R.V. Pound and collaborators, published in 1960. But the frequency of this light beam is also a kind of clock. Thus the principle of equivalence

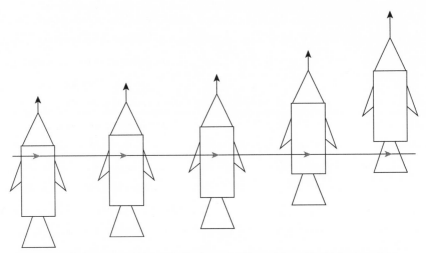

View of the accelerating elevator in which the "being" has been replaced by a rocket propelling the elevater upwards with in acceleration g. The beam of light travels in a straight line; it is the elevator that is accelerating. We can thus imagine that if we were standing in the elevator, the beam of light would thus appear to follow a curved path, as shown below (lower left).

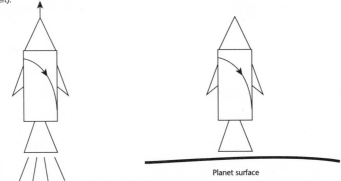

Planet surface

Due to the "equivalence principle," if you were to stand inside the elevator, it would not be possible to tell whether you were accelerating (above left) or whether you were instead placed in a gravitational field, on a planet's surface (above right). Also, because we know that in an accelerating frame like that in the elevator on the left, a beam of light would appear to follow a bent path, we ought to observe the same bending of light if we were on planet's surface, in the gravitational field.

Figure 2.16. The equivalence principle illustrated by the accelerating elevator.

demonstrates that gravitation also alters time as well as space. It was this effect that Einstein had not included when he discussed clocks at the North Pole and the Equator in his 1905 paper. It is what cancels out the special relativity time dilation. Einstein spent the four years after he published his equivalence paper doing the hardest work of his life. The equivalence principle applies only to uniform gravitational fields. Even the gravitational field at the surface of the Earth is not really uniform. In his general theory of relativity, Einstein allowed any kind of gravitational field. At each point in space-time there is, in principle, a different geometry depending on the gravitational fields present. The short hand way of saying this is that gravitation "curves" space-time. Objects follow the equivalent of the straight lines in these geometries. In 1916, Einstein published perhaps the most remarkable physics paper ever written, "The Foundation of the General Theory of Relativity." In it he had to invent, or resurrect, a whole new mathematics. We are still living off the riches of this paper.

One of the predictions of this paper was that the Sun would bend starlight, but through twice the angle his 1911 paper had predicted. This was first confirmed by two British expeditions in 1919. Sir Arthur Eddington, who was a Quaker, and thought that such an international scientific effort might help to bring reconciliation among the warring countries, was the leader of one of the expeditions. He sent a telegram to Einstein congratulating him on the agreement of his theory with their measurement. As it happened, soon after it came, he was with a student named Ilse Rosenthal-Schneider. Her recollection of the occasion is my favorite Einstein anecdote. She wrote,

> He suddenly interrupted the discussion . . . reached for a telegram that was lying on the windowsill, and handed it to me with the words, "Here, this will perhaps interest you." It was Eddington's cable with the results of the eclipse expedition. When I was giving expression to my joy that the result coincided with his calculations, he said quite unmoved, "But I knew that the theory is correct"; and when I asked what if there had been no confirmation of his prediction, he countered, *"Da könnt' mir halt der Liebe Gott leid tun, die Theorie stimmt doch"* (Then I would have been sorry for the dear Lord–the theory is corrrect).

We now leave relativity and turn to Einstein's other 1905 papers.

3
Do Atoms Exist?

⌇ SETTING THE STAGE

The question whether atoms exist or not has but little significance
from a chemical point of view; its discussion belongs rather to metaphysics. In chemistry, we have only to decide whether the assumption of atoms is an hypothesis adopted to the explanation of chemical
phenomena. From a philosophical point of view, I do not believe in
the actual existence of atoms, taking the word in its literal significance of indivisible particles of matter — I rather expect that we shall
some day find for what we now call atoms a mathematico-mechanical
explanation which will render an account of atomic weight, of atom-
icity, and of numerous other properties of the so-called atoms.

—F.A. Kekulé

What made the greatest impression upon the student, however, was less the technical construction of mechanics or the solutions to complicated problems, than the achievements of mechanics in areas which apparently had nothing to do with mechanics: the mechanical theory of light, which conceived of light as a wave-motion of a quasi-rigid elastic ether, and above all the kinetic theory of gases: the independence of the specific heat of monatomic gasses of the atomic weight, the derivation of the equation of state of a gas and its relation to the specific heat, the kinetic theory of the dissociation of gases, and above all, the quantitative connection of viscosity, heat-conduction, and diffusion of gases, which also furnished the absolute magnitude of the atom. These results supported at the same time mechanics as the foundation of physics and of the atomic hypothesis, which latter was already firmly anchored in chemistry. However in chemistry only the ratios of atomic masses played any rôle, not their absolute magnitudes, so that atomic theory could be viewed more as a visualizing symbol than as knowledge concerning the factual construction of matter.

—Albert Einstein

Some decades ago there was a noted philosopher Morris Raphael Cohen who taught at what was then known as the City College of New York. The students were quite remarkable; future Nobel Prize winners, mayors, legislators, and political scientists of every stripe. They had no inhibitions about asking questions in class and their teachers had no inhibitions in answering. In one of Cohen's classes a student interrupted and asked "Professor Cohen, how do I know that I exist?" Without missing a beat Cohen replied, "And who's asking?" When I first heard this all those years ago, I thought that it was very funny, and I still do. But I saw the point. If this student discovered that in a great variety of independent encounters people addressed him by the same first name and often asked after his only sister, he would probably decide that the notion that this was all part of some chimera located in the mind of the Great Over-Soul was, while logically possible, an unnecessary hypothesis. You might as well admit that you exist and get on to something else. I am going to show you that something rather like this occurred in the beginning of the twentieth century when it became generally accepted, after decades of controversy, that atoms do exist. They are not simply "visualizing symbols" for chemists. You will also see that Einstein was, once again, a central figure.

The Greek word $\tau o\mu o\sigma$–"tomos"–means "cut" and when preceded by an α to make $\alpha\tau o\mu o\sigma$ means "without a cut" or "indivisible." This was the Greek atomic idea. Matter could be divided until one reached the $\alpha\tau o\mu o\sigma$ which were the ultimate constituents. Ultimately all that existed were the atoms and the void. Sometime in the first half of the first century BC–dates are uncertain–The Latin poet Titus Lucretius Carus–"Lucretius"–put this philosophy into an epic poem called *"De rerum natura"(On the Nature of Things)*,–which was based on the ideas of the earlier atomists such as Epicurus–from whom the title came. The atoms of the poem come in various sizes and shapes but have in common that they were indivisible. They were also not directly observable with our senses– only their cumulative effect. As an example he describes particles that are seen "dancing in a sunbeam" noting that "their dancing is an actual indication of underlying movements of matter that are hidden from our sight."

I think that it is fair to say that nothing of scientific interest occurred in atomic theory for the next seventeen hundred years. Newton seems to have been an atomist. In his semipopular book *Opticks*, written in English, unlike the *Principia* which was written in Latin, he notes that,

> It seems probable to me, that God in the beginning form'd matter in solid, massy, hard, impenetrable, Particles of such Sizes and Figures, and with such other Properties and in such Proportion to Space, as most conduced to the End for which he form'd them; and that these primitive Particles being Solids, are incomparably harder than any porous Bodies compounded of them; even so very hard, as never to wear or break into pieces; no ordinary power being able to divide what God himself made one in the first Creation.

Robert Boyle, who was a contemporary of Newton, and also an atomist, confirmed with careful experiments what became known as Boyle's Law; the proposition that at a fixed temperature the product of the pressure and the volume of a gas is a constant. If you increase the volume you will proportionately decrease the pressure. But the truly singular work on atomism was done by Daniel Bernoulli. The Bernouillis, who were originally Dutch, migrated to Switzerland where Daniel was born in 1700. Mathematical brilliance seemed to run in the Bernoulli blood. They all were mathematicians: father, sons, uncles, the lot. But Daniel was certainly the greatest genius among them which did not sit well with his father Johann, who often tried to take credit for work his son had done. Daniel Bernoulli was not only a mathematician. He studied medicine and published papers on the mechanics of breathing among other things. In 1725 he went to St. Petersberg with one of his brothers, Nikolaus, who was, of course, a mathematician. Nikolaus died suddenly and Bernoulli then did everything he could to find a position in Switzerland. In 1733, he went to Basel where he obtained a chair in anatomy and botany. There he remained until his death in 1782.

His most famous work was a treatise he called *Hydrodynamica* or "hydrodynamics"–a term he invented–which was published in 1738. In the final version there are thirteen chapters mostly devoted to the behavior of liquids. For example, he demonstrated that a stream of water radially contracts. The tenth chapter, which is what concerns us, deals with what Bernoulli referred to as "elastic fluids," by which he meant gasses. The questions that he discusses seem natural to us, but no one had thought to look at things this way before and no one afterwards for a century until people like Maxwell reopened the subject. Bernoulli accepted the atomic hypothesis. As he noted, a gas consists of "very minute corpuscles which are driven hither and thither with a very rapid motion." His concern was how to use this model to derive the macroscopic properties of gasses that we observe, properties such as Boyle's law. First he had to state what pressure was from this point of view. He imagined that the gas was confined to a cylindrical vessel with a moveable piston at the top. Then he noted that the piston will be held up "by repeated impacts" on it of the "corpuscles." In short, the pressure is produced by the force imparted by vast number of collisions of the molecules of the gas with the surface of the piston. With this picture in mind he was able to reproduce Boyle's law. He also understood that if the temperature of the gas were to be increased the speed of the molecules would increase and that this would increase the pressure if the volume was held constant. Some of his argument is difficult to follow but, if rewritten in terms of modern notation, it is something that you could present to a physics class. What I find so odd about this, is that it stands alone. It came from nowhere and went nowhere for a century. I thought of an analogy. It would be as if the Malla kings of Nepal had erected a skyscraper in Medieval Kathmandu which had both glass windows and air-conditioning.

It is also strange to me that neither Bernoulli, nor anyone else for decades, thought to ask how big these corpuscles were, what was their mass, and how many of them there were in some standard volume of gas or liquid. To these people they were just "corpuscles." The rest of this chapter will be devoted, in one way or another, to these questions. But I want to begin with an instructive example of someone else who failed to ask–in this case Benjamin Franklin. At the time–the late 1700s–Franklin was living near Clapham Common in London. He was in correspondence with a physician and noted scientific amateur named William Brownrigg. It seems that Brownrigg had written him about someone who had observed that pouring oil on troubled waters has a calming effect. In November of 1773, Franklin responded describing his own observations on the matter. Among other things, he had put a small amount of what must have been olive oil on the large pond in the common. He speaks of using a teaspoon of oil. He noted that it spread very rapidly. Indeed, the tea spoon of oil soon covered a half acre and then stopped spreading. He also noticed that until the oil had spread thin it produced what Franklin called "prismatic colors." It didn't seem to occur to him to ask why the oil stopped spreading and why the prismatic colors disappeared. The answer to both of these questions is the same.

The oil spreads until it reached a thickness of the size of a single molecule–a monomolecular layer. It cannot get any thinner than that.

The "prismatic colors" are a little more subtle. If you have a thin layer of, say oil, spread on a substance like water, and shine light on it, what happens is that some of the light is reflected from the surface of the oil and some having penetrated the layer is reflected by the water. The latter follow a longer path and can get out of phase with the former. The two wave forms will be displaced since one wave will arrive after the other. This means that the maxima or minima of the wave will no longer overlap. If the thickness of the layer is such that the difference of path length is a whole number of wave lengths of the light, then the two beams are in phase and reenforce each other. If the difference is say a half wave length then the two beams are exactly out of phase and interfere destructively with each other. Sunlight, as we know, consists of many wave lengths. Looking at a given angle at the layer of oil will pick out that part of the spectrum in which the beams are in phase. If we change the angle then the path lengths change and another color will be picked out–thus the "prismatic colors." However if the layer of oil has a thickness that is much less than the wave length of light there is no effect because the path lengths are sensibly the same. The "prismatic colors" disappear. A useful unit of length in this business is the "Angstrom." One Angstrom — $1\text{Å} = 1/10^8$ centimeters $= 10^{-8}$cm $= 1/100,000,000$ cm. Visible light is in the 4000 to 7500 Angstrom range so we can conclude that molecular sizes must be a great deal smaller than 4000 Angstroms.

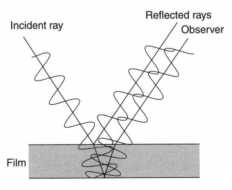

Figure 3.1. Why a thin film produces prismatic colors.

We can do better than that if we pursue Franklin's experiment a little further. To do this we shall make a molecular model. Molecules don't really look like our model but using it we should be able to get correct orders of magnitude. We shall suppose that each molecule is a tiny cube whose sides have length L. I am not sure of the volume of Franklin's tea spoon, but an American teaspoon has a volume which is given conventionally as 5 cm^3. A British tea spoon is a little larger. Thus there are a total number of molecules of oil, N, of about 5 cm$^3/L^3$. We are thinking of the little cubes being stacked up to fill the teaspoon. These same N molecules cover a half acre which is about 2×10^7 cm^2.

Thus we have two equations

$$N = 5\,\text{cm}^3/L^3$$

and

$$N = 2 \times 10^7\,\text{cm}^2/L^2$$

So dividing, we have that L is about 25 Angstroms. This is a large molecule but still incredibly small by any ordinary measure. We may ask about how many of them are in a cubic centimeter in the teaspoon. Using the numbers above we find about 10^{20}/cm^3, a huge number. Shortly I will give numbers that have a more universal character.

As I said, this experimental reasoning was not carried out by Franklin, but a few decades later the first experiments were performed explicitly to find the size of a molecule. They were carried out by Thomas Young, the British polymath whom we met in Chapter I when discussing the evidence for the wave nature of light. In 1816, Young published an article in the fourth edition of the Encyclopaedia Britannica on what he called "cohesion." He was studying the cohesive force between water molecules and determined that the range of the force was about an Angstrom. He conjectured that the size of the molecules must be less than this but could not give a precise value. Actually the size is about two Angstroms. Here matters remained for the next half century. In the meanwhile the chemists had gotten into the act.

John Dalton, who was born in 1766, was the second son of a modest Cumberland weaver. Dalton might have spent his life as a day laborer if his brother had not found him a place in a newly established Quaker boarding school. It was here that Dalton developed his interest in science. He began his scientific career as an atomist, but for reasons I will explain, developed doubts. He envisioned chemical reactions as taking place as combinations of individual atoms which accounted for the fact that elements seemed to combine in definite proportions. In his great book, *New System of Chemical Philosophy*, which was first published in 1808, he gave his rules for chemical combination. Inevitably he got some things wrong. For example, he claimed that the reaction that produces water was a combination of one hydrogen atom with one oxygen atom to produce a water molecule of the form HO. The year after Dalton published his book, the French chemist Joseph Louis Gay-Lussac published his results on how volumes of gasses combine in simple integer proportions. In particular, he found that two volumes of hydrogen combined with one volume of oxygen to give two volumes of water vapor. This was incompatible with Dalton's formula for water and caused him to have doubts about the entire atomic theory of chemistry. In science it sometimes happens that a genius appears just when he, or she, is needed. In this case it was the Italian aristocrat Lorenzo Romano Amedeo Carlo Avogadro di Quarenga e di Cerreto-Amedeo. Avogadro for short.

Avogadro, who was born in Turin in 1776, was headed for a career in the law but after a few years abandoned it for the study of physics and mathematics. By 1811, he had made two of the most important discoveries in the history of chemistry, which he published in the *Journal de Physique*. He recognized that Gay-Lussac's observations on combining volumes of gasses could be explained if equal volumes of these gasses had equal numbers of fundamental particles. But then how to explain the results on water? For this Avogadro made the profound observation that these fundamental particles need not consist of a single atom.

Indeed, both hydrogen and oxygen in this reaction are diatomic. The basic unit in the gas–the molecule–we write as H_2 and O_2 respectively. There are two atoms of each gas per molecule. The claim is that equal volumes of gasses contain the same number of **molecules**. If we recognize that the correct chemical formula for water is H_2O, everything fits. One volume of O_2 plus two volumes of H_2 yields two volumes of H_2O. It is rather amusing that the number of molecules in a cubic centimeter of a gas under standard conditions of pressure and temperature–14.7 pounds per square inch of pressure, one atmosphere, and $32°F$–is usually not now called the Avogadro number, but rather the Loschmidt number, which is incidentally is approximately 2.67×10^{19} molecules per cubic centimeter. Loschmidt will come into our story next. What is generally called the Avogadro number refers to the number of molecules in what is called a gram molecule or "mole." By definition, the mole of any substance weighs in grams an amount equal to its molecular weight. For example, the molecular weight of helium is 4, so a mole weighs four grams. The molecular weight of diatomic oxygen is thirty two, so a mole of oxygen gas would weigh thirty two grams and so on.[1] The number of molecules in a mole is a universal number and is measured to be 6.022×10^{23}. Avogadro never considered this quantity and the name entered physics only early in the twentieth century. By the way, we can use Avogadro's number to find the mass in grams of, say, the hydrogen atom. We can take the atomic weight of hydrogen to be approximately one. If we call the mass of the hydrogen atom m_H and Avogadro's number N_A then $m_H \times N_A = 1$ so that solving we find that $m_H = 1.6 \times 10^{-24}$ grams.

There are a number of appealing characters in our story, but one of my favorites is Jan Josef Loschmidt. Loschmidt was born into a poor farming family, in what is now the Czech Republic, in 1821. If his abilities

---

[1] Atomic weights–the relative weights of elements–are generally nearly integers. Strictly speaking they are defined by finding the atomic weights of the various isotopes of the atom and averaging, taking into account the abundance of the isotopes. We are ignoring this subtlety here.

had not been recognized by a local priest, who persuaded Loschmidt's family to allow him to have an education, no doubt he too would have remained on the farm. Loschmidt went to high school and then for two years to the Charles University in Prague where he studied philosophy and mathematics. He then moved to Vienna where he supported himself as a private tutor while taking the equivalent of a bachelor's degree in physics and chemistry. Then he needed a job. He looked everywhere, including the newly created state of Texas. If Loschmidt had actually gone to Texas the history of chemistry would very likely have been quite different. In 1856, after some unsuccessful business enterprises, he got a job teaching in a high school in Vienna where he was allowed to have a small personal laboratory. It was as a high school teacher that in 1861, he published, at his own expense, a booklet containing his first two papers on chemistry. In them he gave the first formulae for the chemical structure of hydrocarbons. Kekulé, who never acknowledged Loschmidt's work, became famous for his ring structure of the benzene molecule which had been used previously by Loschmidt in a hundred hydrocarbons. In the meanwhile, Loschmidt had become known to people at the University and in 1868, he was made an associate professor of physical chemistry and was given an honorary PhD. But it was in 1865, that he published the paper "On the Size of Air Molecules" in the Proceedings of the Academy of Science of Vienna to which we will now turn our attention.

From the references in his paper it is clear that Loschmidt was familiar with the work of Maxwell and the German physicist Rudolf Clausius on statistical mechanics. There is no reference to Young, so Loschmidt seems unaware of his measurements of the size of molecules. In his own analysis Loschmidt makes use of two results. The first is an analysis by Maxwell as to why, for example, air, offers resistance to objects projected into it—why air is "viscous." Maxwell argued that it came about because of collisions between faster and slower moving molecules. Thus the essential determining factor was the frequency of these collisions. Equivalently we can ask, on the average, what is the distance traveled by a molecule before it collides with another. This distance is referred to as the "mean free path." It is clear that it depends on two things; how many molecules per centimeter cubed are encountered, and the area that each molecule presents for a collision. In fact, the bigger these numbers are, the smaller the mean free path. Ignoring all numerical factors–Loschmidt does it more carefully–I will take the area to be L^2 where, L is the size of the molecule. The number of molecules per cubic centimeter I will call, N_L–the Loschmidt number. If I call the mean free path, l, and ignore all the numerical factors then

$$l = \frac{1}{N_L L^2},$$

which says that the mean free path between collisions decreases with the number of molecules available and the area cut out by each molecule. The measured mean free path that Loschmidt got from the work of Maxwell, and others, was about 0.000014 cm $= 1.4 \times 10^{-5}$cm. We can be sure that the size of the molecule is substantially smaller than this. Our equation for N_L involves two unknowns, L and l, so we need another equation.[2]

To find the second equation Loschmidt reasoned essentially as follows. Suppose we could liquefy air. We could then think of the molecules, like we did for the one's in Franklin's spoon, as being packed together with the little cubes lying on top of each other. Each tube has a volume of L^3 so if we call the mass of the air molecule, m, then the density of the liquid air, ρ_L, again ignoring all geometric factors, is given by $\rho_L = \frac{m}{L^3}$. The density of ordinary air–that is, gaseous air–is given by $\rho_A = mN_L$. In a gas, the molecules are not packed together so the density is just given

---

[2]The existence of such a mean free path explains an apparent paradox in the kinetic theory. Molecules in a gas have average speeds of several hundred meters a second. Thus one might expect that introducing a gas in the corner of a room would be noticed throughout the room in seconds rather than minutes. The reason this delay happens is that the molecules are constantly colliding. Multiplying the time between collisions by this average speed we get the mean free path. Solving for the time we get about a billionth of a second implying that the molecules collide billions of times a second.

by the mass of a molecule multiplied by the number of them in a cubic centimeter.

Thus, dividing the two expressions,

$$\frac{\rho_L}{\rho_A} = \frac{1}{N_L L^3}.$$

Now we have two equations and can first solve them for L. Thus

$$L = l \times \frac{\rho_L}{\rho_A}.$$

Loschmidt used somewhat different quantities and took spherical molecules, but this is the idea. At this point he was somewhat stymied because air had never been liquefied. Indeed its components such as oxygen, argon and nitrogen hadn't been liquefied either. That would come at the end of the century. So he used some indirect chemical arguments to find what he needed to know. We can cheat and use the known values $\rho_A = 1.25 \times 10^{-3} g/cm^3$ and $\rho_L = 880 \times 10^{-3} g/cm^3$. Thus, according to this estimate, L is about 2×10^{-8} cm $= 2$Å. With his spherical air molecule, Loschmidt found the radius of the sphere to be about twice this size. The presently accepted value of the diameter of the sphere is about 3×10^{-8} cm, that is, approximately 3Å. Molecules are not really spheres but this gives a feeling for their size. Curiously, Loschmidt did not use the same reasoning to find N_L which is given in our simplified form by

$$N_L = \frac{\rho_G}{\rho_L} \times \frac{1}{L^3}.$$

If we use Loschmidt's value for L we find an answer for N_L which is an order of magnitude too large. Later Loschmidt did give some estimates for N_L, but these had a range of values. It would take a new generation of experiments in the next century to arrive at the presently accepted value.

From the way I have presented this discussion you might easily get the impression that during this period atomic science was proceeding normally. It was not. There were pitched battles being fought between atomists and nonatomists as to whether atoms even existed or whether they were a burdensome theoretical construct. Chemists disputed with chemists, physicists with physicists and the two groups disputed with each other. A perfect microcosm of this is what happened in Loschmidt's own university, although the principals were not assembled under one roof until after Loschmidt's death in 1895. One of Loschmidt's students was Ludwig Boltzmann who became one of the great theoretical physicists of his generation. Loschmidt presented him with a very deep puzzle. In the statistical mechanical picture of gasses the molecules are colliding with each other billions of time a second. It was by analyzing the average behavior of these collisions that Maxwell, and others, were able to derive some of the laws of thermodynamics such as Boyle's law. But these collisions are all reversible. If you take a given collision what happens is that two molecules each with its individual initial momentum collide. Following the collisions, the two molecules emerge with different momenta. But we can imagine running the collision backwards. The two final momenta become the initial momenta, and the two initial momenta become the final momenta. In one of these gasses this collision is just as likely to occur as the one we started with. This implies the microscopic reversibility of the system. But then how does the system as whole evolve? To take a specific example, suppose we have a container of water and put some powdered dye in a corner. We know what will happen. The dye will diffuse throughout the water and eventually color the whole container. But how can the microscopically reversible collisions account for this? Why does the dye not return to its initial corner? This was a problem that Boltzmann spent much of his professional life resolving. In essence, it comes down to less probable configurations evolving into more probable ones until the system reaches a stasis—an equilibrium. The probability of its reverting to its unlikely initial state is infinitesimal. In his analysis Boltzmann treated the atoms and molecules as real entities.

His counterpart was Ernst Mach, the same Mach who influenced Einstein. Nonhistorians of science largely remember him now because of the Mach numbers, which give the ratio of the speed of a supersonic object to the speed of sound. In 1886, he photographed the shock waves generated by supersonic bullets. Mach spent his early career teaching in a provincial Austrian university, followed as we have noted by several years at the so-called German University in Prague. He returned to the University of Vienna in 1895–the year of Loschmidt's death, but as a professor of philosophy. Despite his fairly undistinguished record as a scientist Mach became on of the most influential physicists of his era. This had to do with his philosophical and historical writing and above all with his masterpiece *The Science of Mechanics*.

In *The Science of Mechanics*, as we have seen, Mach exposes the often unstated metaphysical and even theological assumptions that underlay, for example, the mechanics of Newton. It was, as I have mentioned, Mach's skepticism that helped to inspire Einstein to have the courage to overthrow the Newtonian world view. But, even Einstein, felt that Mach went too far. Mach was a positivist who believed that theories in physics should be nothing more or less than the economical description of observed facts. For Mach, Boyle's law was a model of such a description. It was a relationship between observed quantities such as the temperature, volume, and pressure of a gas. What Mach objected to was the introduction of what he regarded as the extraneous complication of atoms and molecules to "explain" the law. Given his view, and the fact that he and Boltzmann became colleagues, a collision was inevitable. Boltzmann later wrote, "I once engaged in a lively debate on the value of atomic theories with a group of academicians, including Hofrat Professor Mach, right on the floor of the academy [of science] itself... Suddenly Mach spoke out from the group and laconically said: 'I don't believe that atoms exist.' This sentence went round and round in my head." When the subject of atoms came up, Mach was fond of asking "Have you seen one?" This is a fair question and at the end of this chapter we shall return to it when we briefly consider the permanently confined quarks. Now, we are going to have a mathematical interlude, then on to Einstein. It is not an idle interlude because without it we cannot understand Einstein.

MATHEMATICAL INTERLUDE: THE DRUNKARD'S WALK

When I was a kid I used occasionally to be taken to see horror movies, the kind that starred Boris Karloff or Bela Lugosi. When it came to the scary parts I would hide my eyes with my hands. In the really scary parts I would close my eyes and put my fingers into my ears. For some of you a mathematical interlude may seem like an experience of just this kind. If you want, you can close your eyes and put your fingers in your ears until its over. I think you will miss some of the fun. I also think that you may find the part on Einstein less comprehensible than it should be. In any event I am going to analyze, a little bit, one of the classic problems in mathematics–the drunkard's walk.

At the risk of being sexist I am going to take my drunkard to be male. To make the analysis simpler, without sacrificing anything that is essential, I am going to make his walk one dimensional–along a line. In addition, this is what I am going to assume about the walk. It starts at some origin–under a lamp post, say. At each step the drunkard will flip a coin. If it comes up heads he will take one step to the right and if it comes up tails he will take one step to the left. All the steps will be of the same length–say one foot. If the coin is "fair," the probability is a half for a step to the right and half for a step to the left. Probabilities always add up to one since something is sure to happen. Furthermore, nothing in the previous history of the walk affects the probability of the next step. That is why the walk is "random." The question we want to discuss is what will happen. We can see at the outset that there is no definite answer to this question. There are many possible walks. We want to see what, at least, some of them are.

The first step is very simple. The drunkard can end up at -1 or at $+1$ as depicted in Figure 3.2.

These two outcomes are equally probable. Since there are two possible outcomes each one has a probability of one-half. The next step becomes more interesting. The drunkard can end up in one of three locations, $+2$, 0, or -2, as shown in Figure 3.3.

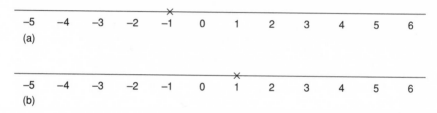

Figure 3.2. In the first step drunkard's walk can have different results, as in (a) −1 or (b) +1.

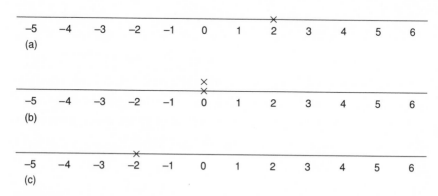

Figure 3.3. In the next step, the drunkard can find himself in one of three locations: (a)+2, (b) 0, or (c) −2.

But note that at the second step there is one way to get to +2 or −2 but two ways to get back to zero. That is what I mean by the two crosses in the figure. Since the total number of possible walks at this step is four, the probabilities of getting to plus or minus two is one-fourth, while the probability of getting back where you started is one-half. You may be comforted to know that there are general formulae for these probabilities, and that I am not going to write them here. They can be applied for an arbitrary number of steps. But there is a geometrical construction that gives these outcomes. It is called *Pascal's triangle* named after the 17th-century French mathematician Blaise Pascal. It seems to have been discovered much earlier by the Chinese. This is what it looks like for the first few steps (Figure 3.4).

We can see the meaning of the triangle by considering a couple of examples. As we have mentioned, in the second step there are two ways of getting back to the center and one way to get to either extreme. This is reflected in the entries 1, 2, 1 in the triangle. Now you

```
                  1
                1   1
              1   2   1
            1   3   3   1
          1   4   6   4   1
        1   5  10  10   5   1
      1   6  15  20  15   6   1
```

Figure 3.4. The first few steps of Pascal's triangle.

can see what the other entries mean. For example, in the sixth step there are twenty different routes that take you back to the center. The total number of possible routes escalates with the increasing number of steps N. For example for $N = 6$ we have, adding them up, 64 possible routes. It is no coincidence that $64 = 2^6$. We can show quite generally that the number of routes at the Nth step in 2^N. Here is another interesting thing about the triangle. Suppose we consider, for example, the polynomial $(a + b)^5 = (a + b)(a + b)(a + b)(a + b)(a + b) = a^5 + 5a^4b + 10a^3b^2 + 10a^2b^3 + b^5$. You will notice that these coefficients are the same numbers as you find in the fifth row of the triangle. This is also not an accident. The coefficients of polynomial expansions can be read off the triangle. I will spare you the proof which would require my writing down the formula for an arbitrary entry in the triangle. Those of you who might be interested can find it on the web. Note also we can find the entries on the next row by adding up adjacent entries on the previous row.

The number of routes escalates as the number of steps increases. So does the complexity of the routes (see Figure 3.5). What is plotted here are the places where the steps take you, versus the number of steps. There is a bit of artistic license here. The steps jump discontinuously, but the in-between points have been filled in. As the number of steps gets larger the in-between points become less significant and to all practical purposes the curves do get filled in. I think it is fair to say that just looking at these curves without explanation, one would be unlikely to come up with the correct explanation. It is remarkable that such a simple rule as the drunkard's walk can generate curves of this complexity.

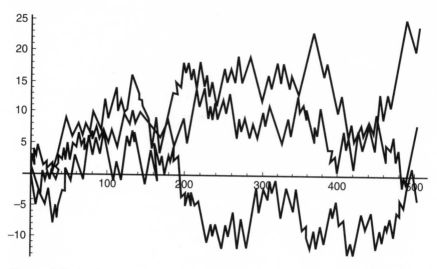

Figure 3.5. As the number of steps increases, so does the complexity of possible routes.

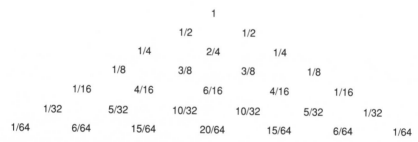

Figure 3.6. The probability triangle.

In the triangle I have given the number of routes available to reach a certain outcome of the walk. It is easy to convert this into a triangle of probabilities. We just have to divide the entries on each row by the total number of entries. For example on the fifth row we have the entries 1, 5, 10, 10, 5, 1 which give the number of routes that end up at distances of 5, 3, 1 or −1, −3−5 respectively. To get the probabilities of ending up at a given place, we divide its number by 32, which is the total number of routes when 5 steps are taken. We can plot these probability triangle numbers as shown in Figure 3.6. The can be converted in a histogram (Figure 3.7) and morphed into a Bell curve (Figure 3.8).

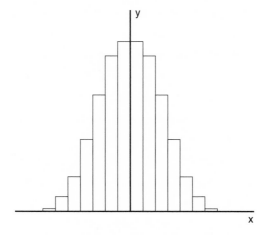

Figure 3.7. The probability triangle converted into a histogram.

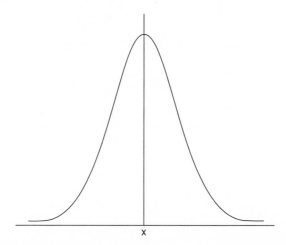

Figure 3.8. The morphing of the histogram into a bell curve as the number of steps increases.

As the number of steps increases the histogram will gradually morph into the bell-shaped curve above. In the limit in which the number of steps becomes infinite the histogram becomes the bell-shaped curve. This is an example of one of the most important theorems in statistics and probability. It is called "the central limit theorem." What it says is that if you have a huge number of random events their probability distribution becomes "normal", or "Gaussian normal," which means that the curve that describes these probabilities becomes bell shaped. That is why the bell-shaped curve is so ubiquitous in statistics.

Now I want to finish this interlude with something that I want you to pay special attention to. I want to introduce a measure of how far on the average the drunkard will travel away from where he started after N steps. We could try to keep track of the positive and negative steps which would be rather complicated. Instead we will perform the average by averaging the squares of the distances which are all positive. This will give us the "mean square distance." Its square root will give the "root mean square distance." This may sound a little complicated, but I think you will be charmed by the answer. I hope so because it is the key to understanding Einstein's results, which we come to shortly. To be concrete I will consider $N = 5$. As mentioned before the drunkard can end up at the distances -5, -3, -1, $+1$, $+3$, $+5$. The total number of routes is 32 so looking at the triangle the probabilities of these outcomes are 1/32, 5/32, 10/32 and 10/32, 5/32, 1/32 respectively. Note that the probabilities add up to one as they are meant to. Now we can form the average, which I will call for the Nth step $\langle x^2 \rangle_N$. This is the average I get by squaring each distance and multiplying by its respective probability. Thus for five steps I get

$$\langle x^2 \rangle_5 = \frac{1}{32} \times (1 \times 25 + 5 \times 9 + 10 \times 1 + 10 \times 1 + 5 \times 9 + 1 \times 25) = 5.$$

To be careful I should put in the unit in which I am measuring distances. If the steps are a foot at a time then the unit that should be there is feet squared. The unit for $\sqrt{\langle x^2 \rangle_5} = \sqrt{5}$ would be in feet. This simple result is very general. If we repeat it for the Nth step we will find that $\sqrt{\langle x^2 \rangle_N} = \sqrt{N}$ which in this case would be measured in feet. One last thing. Let us suppose that to make the N steps took the drunkard a time t. Let us imagine that this time is divided up into units, say Δt, so that $\frac{t}{\Delta t} = N$. The result we have in terms of the time is that the root mean square distance the drunkard goes in a random walk that takes the time t to perform, increases as the square root of the time. In other words, if the drunkard walks for four times the time he will, on the average, only get twice as far. This is essential to understanding Einstein's result.

∽ BROWNIAN MOVEMENT

> In this paper it will be shown that according to the molecular-kinetic theory of heat, bodies of microscopically-visible size suspended in a liquid as a result of thermal molecular motions, will perform movements of such magnitude that they can be easily observed in a microscope. It is possible that the movements to be discussed here are identical with the so-called "Brownian molecular motion"; however, the information available to me regarding the latter is so lacking in precision that I can form no judgment in the matter.
> —Albert Einstein, 1905

Robert Brown was born in Montrose, Scotland, in 1773 (see Figure 3.9). His father was a clergyman. But Brown studied to be a doctor and, in 1795, he joined the Fifeshire Regiment of Fencibles as a surgeon's mate. His diary is still extant. An entry dated 13, January 1800, reads

> At breakfast read part of the rules concerning the gender of German nouns in Wendelborn's grammar. After breakfast transcribed into my botanical common place book part of my notes on Sloane's Herbarium on Jamaican Ferns. Attended the Hospital from one till three o'clock. Drank about a pint of port in negus. Conversation various...About twelve o'clock finished the transcription of my notes on Sir Hans Sloane's Ferns. This transcription has not afforded me one new idea on the subject of Filices.[3]

At the end of 1800, Brown was offered the post of naturalist aboard a sailing ship called *Investigator* for the very large sum of four hundred twenty pounds. This had come about because Brown had met, and impressed, Joseph Banks, an independently wealthy and well-known botanist who had influence with the Admiralty. In the event, the voyage,

---

[3]*Dictionary of Scientific Biography, op.cit.*, Vol.II, p.517. "Negus" was a popular mixed after-dinner drink and "filices" is the Latin term for "ferns."

Figure 3.9. Robert Brown. (Courtesy, Topham/The Image Works)

which went to Australia, lasted four years. Brown came back with a treasure trove of some four thousand new botanical species as well as drawings and zoological specimens. Dealing with this material took the next several years. On Banks' death in 1820, Brown was given an inheritance that included a house and Banks' library and zoological collection. He was very busy with all of this and published very little for several years. During this period Brown honed his skills as a microscopist. Among other things he observed plant cells and identified a part he called the "nucleus." By 1827, he was studying pollen grains when he made the discovery that has immortalized him.

He noticed that suspended in the fluid in the grains were tiny particles. Typically the particles had the size of a few ten thousandths of a centimeter. They were indeed microscopic. They were also undergoing a kind of St. Vitus dance—jigging in various apparently random directions. He came to the conclusion that the movement did not have to do with currents in the fluid, but seemed to belong to the particles themselves. Movements like this had been reported previously but by no one with Brown's scientific curiosity. Among his specimens he had many different kinds of fresh pollen to experiment with, and he noticed that the microscopic particles in all of them underwent the same kind of jittery movement. Incidentally, Brown and his successors referred to this effect as "movement" and not "motion," so in this spirit I will always refer to Brownian "movement." Then he turned to dead pollen, some of which had been preserved in the herbarium for a century. They too had particles that when suspended in water jittered. Next he tried truly inanimate objects. These included London soot, and a finely powdered stone that had been taken from the Sphinx. All of them, when ground into microscopic particles, showed the same movement. In 1828, he issued a privately printed pamphlet with the cumbersome title *A Brief Account of Microscopical Observations Made in the Months of June, July and August of 1827, on the Particles Contained in the Pollen of Plants; and on the General Existence of Active Molecules in Organic and Inorganic Bodies.* Brown was a curious mixture of the social and antisocial. He never married, and lived alone for the rest of his life in the house that Banks had bequeathed him. Nonetheless, he joined various scientific bodies such as the Royal Society. He also spent Sunday mornings with Darwin who had a great admiration for him. Thus, despite the fact that Brown had only published his discovery in a privately printed pamphlet to which he had never added anything, it became rather widely known.

There were a variety of reactions including that of Maxwell, who conjectured that if Brown had used a more powerful microscope the movement would disappear. There have been, incidentally, some modern commentators who have claimed that Brown never could have seen what he claimed to have seen with the microscope at his disposal. But the microscope is still extant and has been used to take moving pictures of Brownian movement. It is quite visible. Figure 3.10 shows the actual microscope.

The common explanation at the time was that the movement was due to small currents in the liquid. By the end of the 19th century, the idea that the motion was caused by the incessant collisions of the larger particles with the invisible molecules of the liquid in which they were suspended, began to take prominence. This, of course, required

Figure 3.10. The microscope used by Brown in his experiments on the movement of suspended particles. (Photo courtesy Brian J. Ford)

the "existence" of the molecules and that became part of the debate. There were objections, some of the most interesting coming from a distinguished German cell biologist named Karl von Nägli. Nägli suggested two difficulties. Firstly, he said that because of the disparity of size between the particles and molecules it would take millions of collisions for the molecules to move the particles significantly. However, he seemed to realize that there were enough collisions per second—quintillions in fact—so that this was not a problem. But his second objection is more interesting. He said that since these collisions are coming from all directions at random, the particles will never be able to move. One collision will be offset by another. Those of you who have read the mathematical interlude will immediately see the flaw in this. We might also have said in that example that since the drunkard at each step has equal probability of going left or right, he will never get anywhere. But we have seen that this is not true. He will move away from where he started from with a root mean square distance that is proportional to the square root of the elapsed time. Nägli's contemporary the French physicist Léon Gouy did a series of experiments that showed, among other things, that Brownian movement took place in a wide variety of solvents and had nothing to do with using water. Gouy insisted that the only possible explanation of his experiments was that motion arose from the collisions of the particles with the invisible molecules of the solvent. Indeed, this phenomenon was a proof of the validity of the kinetic theory. However, neither he nor anyone else, had actually shown how to use the theory to derive the properties of Brownian movement. Enter Albert Einstein.

Einstein's connection with this problem is related to the somewhat baroque story of his PhD thesis. At first sight one might be tempted to ask why on earth did Einstein need a PhD? But one has to put oneself back to 1903. Einstein was then recently married, and had been hired as an "Expert III Class" at the patent office in Bern. He had started a thesis for the Polytechnic in Zurich but had withdrawn it in 1902, when there did not seem to be much enthusiasm from the faculty. At Bern, he decided to take his PhD up again. He definitely wanted to start up the academic ladder. This had been his hope in Zurich. The lowest rung was what was *privatdozent*, which allowed one to teach at a university, earning only the money that students paid for the lectures. Even this required a PhD. Einstein had heard that the University of Bern would, in special cases, forgo the PhD thesis if the candidate could offer the proof of significant original work done prior to the demand for the degree. Einstein tried this in 1903, and was turned down. This may also seem crazy but the papers he submitted were a little strange.

By 1903, Einstein had published four papers, all of them in the German journal *Annalen der Physik*. The first two he later referred to as "beginners' papers" and in retrospect he would not have submitted them. The next two had to do with the foundations of statistical mechanics. They were very likely turned down as a substitute for a thesis for the wrong reason—no one at the University of Bern understood them. The right reason would have been that the work had been done independently by Boltzmann and by the American physicist Josiah Willard Gibbs who actually died in 1903. He summarized his work in a book he published in 1902, but it was not translated into German until 1905. It is unlikely that anyone in Germany knew anything about Gibbs. This certainly included the then editor of the *Annalen*, Paul Drude, who could accept or reject papers submitted to the journal. Drude, as I mentioned in the last chapter, worked in optics and electromagnetism and not statistical mechanics, which is probably why he did not realize that Einstein's work overlapped with what Boltzmann had already done. Einstein later remarked that if he had known of this work there would have been no need to publish his papers. Nonetheless, they gave him a mastery of statistical mechanics that he made use of throughout his scientific career. T.D. Lee, a Chinese-American physicist who won the Nobel Prize in 1957, with his then collaborator C.N. Yang, told me an anecdote about Einstein. At about the time that he and Yang were doing the work in elementary particle physics for which they won the prize, they were also doing fundamental work in statistical mechanics. Yang was at the Institute for Advanced Study in Princeton where Einstein was resident. They decided to see him and discuss with him their work. Lee had not met Einstein before. Two things struck him. The first was Einstein's hands which Lee recalled were large and appeared to be very strong. The second was Einstein's immediate grasp of what they had done. He asked relevant and fundamental questions. Statistical mechanics stayed with him all his life.

When the University of Bern rejected his papers as a substitute for a thesis, Einstein was really angry. He decided he had had enough of academic politics and did not need a PhD anyway. But, by the spring of 1905, he had cooled off and decided to write a thesis for the Polytechnic in Zurich, despite the fact that he had had a hard time with them earlier. He was in the midst of working on the papers that would form the basis of twentieth-century physics, but he chose for his thesis topic something that he thought would not be controversial–the size of molecules dissolved in a solvent. In this he got involved with the physics of osmosis. On hearing the word "osmosis" you may have the image of putting a CD with an Italian lesson on while you are sleeping, and learning Italian by "osmosis." But in physical chemistry–which is what this subject really is–the term has a definite meaning. We can illustrate it by an example–sugar dissolving in water. What happens when sugar dissolves in water is that the water molecules surround the glucose (simple sugar) molecules in the sugar and tear them from the sugar crystals. Thus, when the process is completed, you have a mixture of glucose molecules and water molecules. Einstein studied the thermodynamics of this mixture something that might sound somewhat mundane, but was not.

The effect that came to be called osmosis had been the subject of some important work at the end of the century. On the instrumental side, a German chemist and botanist named Wilhelm Pfeffer created membranes that would allow, for example, the transport of water, but block the transport of the glucose molecules that were larger. The membranes he created were strong enough so that they could withstand fairly high pressures. Membranes, which he knew were the natural analogues of the artificial ones he had made, operated in plant cells. When one of Pfeffer's artificial membranes is inserted in water, and then sugar put on one side of this barrier, what happens is that water will flow from the side with no sugar into the side with the dissolved sugar. More generally, the flow will be toward the side with the highest concentration of solute. If before adding the solute the water levels were the same, then after adding the solute the water level will rise. This produces pressure on the membrane. This pressure is called the osmotic pressure and the process of diffusion is called osmosis. This pressure can be measured. In fact it can be quite large. A one percent solution of sugar can exert a pressure of two thirds of an atmosphere, which is why you need a solid membrane.

Pfeffer did not propose a theory of this pressure. That was the work of the Dutch physical chemist Jacobus van't Hoff, who won the first Nobel Prize in chemistry in 1901. Van't Hoff showed how to connect this pressure to thermodynamic quantities such as the temperature of the solution and the volumes involved. He had been directed to Pfeffer's work by a plant physiologist at the University of Amsterdam named Hugo de Vries who was looking for an explanation of it. Van't Hoff reasoned that the source of the osmotic pressure was the same as an ordinary gas pressure; namely the glucose molecules were colliding with the membrane and transferring momentum. Both he, and the Nobel committee, took the existence of these molecules as being self-evident. Van't Hoff further reasoned–or conjectured–that if the solution was fairly dilute, the solute molecules–sugar in the example I mentioned–would obey a perfect gas law. That is, at a fixed volume, the osmotic pressure would be proportional to the temperature with the same constants that held for a gas. This was in essence the state of knowledge that Einstein had available when he began working on his thesis. What Einstein did was to use Van't Hoff's theory in his study of how the dissolved molecules flow in the solvent. I will not try to adumbrate his model except to remark that from it he was able to determine a value of Avogadro's number of molecules in a mole. As it happened, the value he determined was too low. It turned out that he had made a mistake in the algebra that was corrected a few years later, producing a value closer to the correct one. In his thesis he also found a value for the molecular size which was also not quite correct, a consequence of the same algebraic error. He finished the thesis at the end of April but did not submit it until July. He was of course, during this period, otherwise occupied creating twentieth-century physics. The thesis, incidentally, was accepted, but the paper based on it was not published until 1906, although logically it should have preceded the Brownian movement paper.

Einstein's first paper on Brownian movement, published in the *Annalen* in 1905, is for lack of a better word, laconic. Indeed, it was so opaque that he received a request from the chemists to write a simpler version.[4] This he published in 1908. It is certainly simpler than his

---

[4] It is amusing that in the translation of this paper found in Furth, *op.cit.*, the individual identified as having made this request is given as R. Lorentz [sic]

1905 paper, but the mechanism that causes the motion, which Einstein refers to as "the heat content of a substance," is not exhibited explicitly. From what we presented in our mathematical interlude we know what the mechanism is. The random bombardments of the molecules of the solute by the molecules of the solvent cause the former to undergo a drunkard's walk. This is the Brownian movement. It was presented in just this way by Einstein's contemporary the Polish theoretical physicist Marian Smoluchowski. Smoluchowski informs us that he had started his work around the year 1900, in response to several papers on the subject. Indeed, Smoluchowski's paper, unlike Einstein's, refers to this work. There is even a refutation of the argument of Nägli about the Brownian particle not being able to get anywhere, along the lines I have mentioned. But Smoluchowski did not publish anything until after he had seen Einstein's paper. In our profession priority goes to those who publish. Smoluchowski recognized this. I think it is also fair to say that while Einstein's paper was more opaque it was also more profound.

In the first part of the paper Einstein is concerned with a thermodynamic question. His work on osmosis had been based on van't Hoff's observation that if the molecules in solution are not too dense then thermodynamically they behave like a gas. What "not too dense" means is that the interactions among the molecules in the solution can be ignored and only the interactions between molecules in the solution and those of the solvent need be considered. The question Einstein asked is why should it be any different for the Brownian particles? They are just larger than, say, the glucose molecules in a water solution. What he showed was that there was no difference and van't Hoff's laws should apply here as well. He then presented two arguments that exhibited how the Brownian particles should diffuse. The first is really a reprise of his thesis. In the fourth part of his paper Einstein gives his second argument. In the course of it he matter of factly introduces a probability

---

which might give the impression that it was H.A. Lorentz himself that made the request. It surely came from the Austrian physical chemist Richard Lorenz.

method that mathematicians and theoretical physicists have expanded on ever since. What he does is to derive an equation from which one can learn how the Brownian particles diffuse, given a rule that describes the likelihood of changes in the positions of individual particles.[5] We used such a rule in the drunkard's walk when we stipulated that it was equally likely for the drunkard at any stage to take a step to the left or right. Here is a small historical footnote. Unknown to Einstein, the same equation had been derived five years earlier in a PhD dissertation by a young French mathematician named Louis Bachelier. In the same thesis Bachelier also did the problem in the way Smoluchowski did, thus anticipating both Einstein and Smoluchowski. Bachelier was not interested in the physics of Brownian movement which he does not mention in his thesis, but rather in the fluctuations of the French stock market–the bourse. He is now recognized as the father of financial engineering. As it happened, Henri Poincaré was one of the examiners on the thesis and wrote the report. Poincaré was also familiar with the work on Brownian movement. It seems odd that when he and Einstein met at the Solvay Congress in Brussels in 1911 for the first time, where Brownian movement was discussed, he did not tell Einstein about Bachelier, who was only rediscovered in the 1950s. However, we know from Einstein's correspondence about the conference that he felt that Poincaré did not understand the new physics such as the quantum theory, and was disappointed by him. It would not be surprising, given Einstein's nature, that he managed to communicate, in one way or another, these feelings to Poincaré.

Those of you who read my little mathematical interlude may have a question. In the interlude I dealt with a one-dimensional random walk. The drunkard could move at each step either to the left or right with equal probability. But here we have a three-dimensional situation. At every

---

[5]Technically what Einstein did was to present a version what later became known as a Chapman-Kolmogorov relation. Both Chapman and Kolmogorov acknowledged that were generalizing and rigorizing Einstein's work.

stage the suspended particle can move in any direction in space. At first this would seem to hopelessly complicate the problem. But it is the very randomness that saves us. Let us call the three directions, x, y, z. At each step the particle will be moved by the molecular collisions in a random direction. This implies that the mean square average distances in each of these directions must be the same. In formulae $\langle x^2 \rangle = \langle y^2 \rangle = \langle z^2 \rangle$. Now we can imagine the following situation. We start the suspended particle off at, say, time zero at some point in the liquid. We draw a set of spherical surfaces, one inside the other, around this point. In due course our particle will cross these surfaces. We cannot predict where on the surface it will cross but we can, as I will now argue, predict at a given time on which surface the crossing will take place. Let is call the radius of one of these spheres, R. Then all the points on the surface satisfy the equation $x^2 + y^2 + z^2 = R^2$. Now if we take the averages and put in the time explicitly we have $\langle x^2(t) \rangle + \langle y^2(t) \rangle + \langle z^2(t) \rangle = \langle R^2(t) \rangle$. So if we use the result above and arbitrarily pick out the x direction we have $3\langle x(t)^2 \rangle = \langle R(t)^2 \rangle$. This tells us that we can use our results from the one-dimensional case.

Einstein, of course, understood this. He does his analysis in one dimension and then applies this result. We have argued that $\sqrt{\langle x^2(t) \rangle} \approx \sqrt{t}$. But Einstein is at great pains to find the coefficient that multiplies the square root. He finds that this coefficient increases with the temperature and decreases with the size of Avogadro's number. This is plausible. If you increase the temperature you will increase the agitation of the molecules that cause the Brownian movement. If you increase the number of molecules that the Brownian particles encounter, you will increase the viscosity of the solvent. Einstein gives an estimate of how far a Brownian particle would go in a minute–what the radius of the spherical surface mentioned above would be if you took water at seventeen degrees centigrade as the solvent. He puts in an approximate value of Avogadro's constant-6×10^{23} particles per mole. He takes the size of the Brownian particle to be about a thousandth of a millimeter. He finds that, in a minute, the average distance traveled by such a Brownian particle would be about a thousandth of centimeter–visible only under a microscope. He ends the paper by saying that "It is to be hoped that some inquirer may succeed shortly in solving the problem [of measuring the motion] suggested here, which is so important in connection with the theory of heat." Later Einstein confessed that he did not think it would be possible for any experimenter to do precise experiments on Brownian movement. He did not reckon with the French physicist Jean Perrin.

Perrin was born in Lille in 1870. He died in New York City in 1942. He had fled France for the United States after the German invasion. He came from a very modest background. His father had been a professional soldier who had been killed when Perrin was still a child. His mother raised him, along with his two sisters. He was very fortunate because the French educational system offered advanced educational possibilities to gifted students, whatever their background. He was accepted in 1890 to the École Normale Supérieure–one of the *grandes écoles*. Here he received a first class education in physics, surrounded by equally gifted students. The existence of molecules was still a much debated topic. Perrin's teachers were strongly in the camp that accepted their reality. Perrin decided that he would devote his career to proving this. He first did work on x-rays but, after he was appointed to teach at the Sorbonne, he gave a course in what was then the new subject of physical chemistry. It was in preparing for this course that Perrin reviewed all the pre-Einstein work on Brownian movement. By the time Einstein's paper appeared, Perrin was already considering how to do precision experiments. By 1908, he had done experiments in which he could map out the details of the jittery paths of individual molecules. When one looks at his results one is struck by how much they resemble our drunkard's walk (Figure 3.11). He could control the external parameters of his experimental set up well enough to get an excellent value for Avagadro's constant. By this time there were several values all done by different methods and all in essential agreement. Perrin's work was generally regarded as remarkable and in 1926, he was awarded the Nobel Prize in physics.

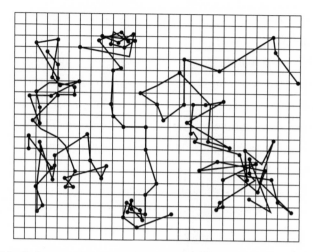

Figure 3.11. A diagram taken from Perrin's Notebooks.

It is fair to say that after Perrin's experiments most of the opposition to the reality of atoms disappeared. An exception was Mach. Near the end of his life–he died in 1916–Mach wrote a preface to his book on physical optics in which he remarks, "But I must assuredly disdain to be a forerunner of the relativists as I withhold from the atomistic belief of the present day." By this time no one much cared. What of the present? How would a present-day physicist answer Mach's question, "Have you seen one?" Probably most—if not all—would answer "yes." They might have in mind the observations made by what is known as a "field ion microscope" invented in 1951, by Erwin E. Mueller of Penn State. Here is the idea. The specimen to be examined is in the form of a very sharp tip–say tungsten. This is surrounded by a gas, say neon or argon. The gas molecules bounce off the tungsten atoms and in the process lose an electron. This ionizes a gas molecule which is then propelled by an electric field onto a fluorescent screen. I simplify here. What appears on the fluorescent screen is an image of the atoms of, say, the tungsten. Is this seeing the tungsten atoms? Most physicists would say yes. Even more striking is the work of the University of Washington physicist Hans Dehmelt. In the late 1980s he managed to trap single atoms in a magnetic field configuration. The atoms were made to fluoresce by shining lasers on them. You can see them. Does this prove, at long last, that atoms exist? This is a question I leave to the philosophers. I also leave to them the question of whether quarks exist. If present ideas are right, free quarks can never be observed unless somehow we can recreate the conditions of the very early universe. The arguments for them are even more indirect that the ones that involve the field ion microscope. The arguments are indirect but impressive. They never would have convinced Mach, but they convince me.

Finally, there was yet another twist involving a thesis. Having a PhD was necessary but not sufficient to take the first step up the academic ladder. You also needed what was called a *Habilitation*—a kind of license. This required yet another thesis. In 1908, after some reluctance, Einstein submitted an *Habilitation* thesis to the dean at the University of Bern. He also had to give a trial lecture. The subject of his lecture was *On the Limits of Validity of Classical Thermodynamics*. The subject of his thesis was *Conclusions from the Energy Distribution Theorem of Black Body Radiation, Concerning the Constitution of the Radiation*. This had been the subject of the first and most revolutionary of his 1905 papers, and it is the subject of the next, and last, chapter of this book.

4
The Quantum

Einstein's office at the university [the German University in Prague] overlooked a park with beautiful gardens and shady trees. He noticed that there were only women walking about in the morning and men in the afternoon, and that some walked alone sunk in deep meditation and others gathered in groups and engaged in vehement discussions. On inquiring what this strange garden was, he was told that it was a park belonging to the insane asylum of the province of Bohemia. The people walking in the garden were inmates of this institution, harmless patients who did not have to be confined. When I first went to Prague, Einstein showed me this view, and said playfully: "Those are the madmen who do not occupy themselves with the quantum theory."

—Philipp Frank

⤳ ENTROPY

Einstein arrived in Bern in February of 1902. His job at the patent office had not yet materialized so to support himself he decided to give private lessons in physics. He put an ad in the *Anzeiger der Stadt Bern*, the local newspaper, which read

> Private lessons in mathematics and physics for students and pupils is given with thoroughness by Albert Einstein, owner of the Swiss polyt.subject teacher diploma, Gerechtigkeitsgasse 32.1st floor. Trial lessons for free.

He got a few takers, one of whom was a Rumanian student of philosophy named Maurice Solovine. Einstein enjoyed talking to Solovine so much that he suggested that they forget about the lessons and simply talk. Not long after they were joined by a young mathematician named Conrad Habicht whom Einstein had already known. The three of them took to meeting regularly in Einstein's flat. A spartan dinner would be served–sausages and the like–there was not much money–and then they would discuss philosophers like Mach or David Hume or even Don Quixote until early hours in the morning. They decided to call themselves as a joke "Akademie Olympia"–the Olympic Academy. On one of Einstein's birthdays Habicht brought caviar which Einstein had never eaten. However, Einstein began a discussion of Galileo's principle of inertia and ate all the caviar without paying any attention to what he had eaten.

By 1905, the Academicians had gone their separate ways. In May of 1905, Habicht received a letter from Einstein which, every time I read it, fills me with the same wonder. He explains that he has written, or is about to write, four papers,

> . . . the first of which I could send off soon, as I am to receive my free copies very shortly. It deals with radiation and the energetic properties of light and is very revolutionary, as you will see provided you send me *your* paper first. The second paper is a determination of the true size of atoms by way of the diffusion and internal friction of

diluted liquid solutions of neutral substances. [This was Einstein's PhD, thesis which was not published until the following year.] The third proves that, on the assumption of the molecular theory of heat, particles of the order of magnitude of 1/1000 millimeters suspended in liquids must already perform an observable disordered movement, caused by thermal motion. Movements of small inanimate suspended bodies have in fact been observed by physiologists and called by them "Brownian molecular motion." The fourth paper is in the draft stage and is on electrodynamics of moving bodies, applying a modification of the theory of space and time; the purely kinematic part of this paper is certain to interest you.

One barely knows where to begin. Three things strike me at once. First, there is the sheer magnitude of the work involved. Each of these papers has many detailed calculations. He must have been able to perform these with amazing speed and precision. He was, after all, working a full schedule at the patent office and had home responsibilities as well. The second thing that strikes me is his absolutely lucid understanding of what he had done, or was about to do. For example, there is the little phrase "the purely kinematic part of this paper is certain to interest you." "Purely kinematic part" means that part that applies independent of any specified force. It is just here where Einstein separates himself from everyone else. People like Lorentz and Poincaré viewed the problem posed by the Michelson-Morely experiment as a dynamical problem. Lorentz in particular showed how the Michelson-Morely experiment could be "explained" if the electrical forces holding matter together caused a contraction of a moving object like a ruler. As we have seen, these models play no role in Einstein's formulation of relativity. The contraction is a consequence of a modified view of space and time. Finally, there is the characterization of the first paper as being "very revolutionary"—not relativity but radiation. The object of this chapter is to help you to understand why this is so.

Not everyone would do so, but I start the road that finally led to Einstein's 1905 radiation paper, with the invention of the steam engine. The first practical steam engine was invented early in the 18th century by

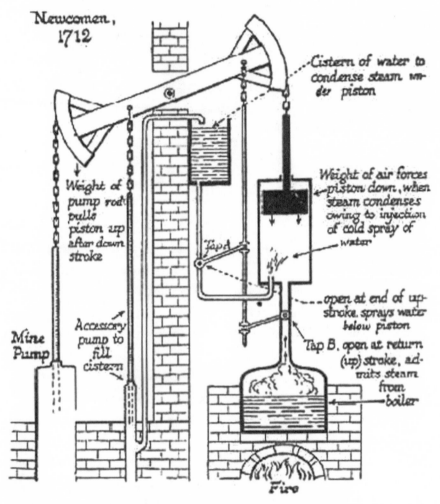

Figure 4.1. Newcomen's Atmospheric Fire Engine.

a British blacksmith named Thomas Newcomen. Like all such engines it was designed to take advantage of the motive power of steam (Figure 4.1). When water is heated enough, it turns into steam. The steam, if introduced into a container, will, like any gas, exert a pressure on the walls of the container. If one of the "walls" is a piston then the steam pressure will drive the piston which can be attached to a lever, and thus perform a useful task. In this case the useful work was done when the heavy lever pushed down on the piston with the force gravity. To do anything interesting, the engine must be able to repeat this cycle again and again. In Newcomen's engine, cool water was introduced into the piston cylinder after the piston had been raised by the steam. This cooling caused the steam to condense creating a vacuum so that the air pressure above the piston would drive it down, and the cycle could begin again. This form of the steam engine was actually used to drain water from coal pits. But, as an engine, it was very inefficient. Part of the reason was that cooling the cylinder meant that in the next cycle the steam remained partially condensed. I have read that the Newcomen engine was less than one percent efficient. More than 99 percent of the steam power did not do useful work. Enter James Watt.

Watt was born in Greenock, Scotland, in 1736. His father was a maker and supplier of nautical instruments. This is what Watt wanted to do as well. In 1775, he went to London to study this trade, but after a year came back to Scotland. The guild in Glasgow would not accept him. He was fortunate to get a job at the University of Glasgow making and repairing scientific instruments. One of these was a model of a Newcomen engine which Watt was asked to repair. He took the opportunity to study the model and realized how it could be transformed. The problem was that there were two apparently contradictory requirements. On the one hand, the cylinder that housed the piston had to be cooled down so the steam would condense and, on the other hand, it had to be kept hot so that on the next cycle the water would remain vaporized. Watt realized that both requirements could be satisfied if the condenser for the steam was in a vessel that was separate from the cylinder that housed the piston. Figure 4.2 is Watt's first drawing of this arrangement.

By 1765, Watt was writing to people that he had invented the "perfect" steam engine. It is true that the Watt steam engine with its many adumbrations, some invented by Watt, became the basis of the industrial revolution, but was it "perfect?" Indeed, what could such a question possibly mean? The answer was supplied by a French engineer–it is difficult to know what else to call him–named Nicolas Léonard Sadi Carnot.

Carnot, who was born in Paris in 1796, was the son of the polymath Lazare Carnot who worked in a variety of things including mathematics for which he is best known. He also played an important role

Figure 4.2. In Watson's first drawing of a steam engine, the piston is at the top and the condenser at the bottom.

in Napoleon's government. His son showed considerable brilliance as a student and at sixteen he entered the École Polytechnique in Paris[1] which trained engineers for the military. When Carnot finished his studies he was commissioned as a second lieutenant doing garrison duty in the provinces. In 1819, he was appointed to the army general staff corps in Paris which gave him time to take courses at the Sorbonne and to do his own research. During the next few years he did the work that would become the basis of the later developments 19th-century thermodynamics. In 1824, he published some of it in a 118-page monograph with the long title, *Reflexions sur la puissance motrice du feu et sur les machines propre à développer cette puissance* which I would translate as "Reflections on the motive power of heat and on the machines that are appropriate to making use of this power." During his lifetime–he died of cholera in 1832 at the age of thirty-six–very few people read his monograph. It was only discovered in mid-century along with some unpublished notes that indicate that Carnot was beginning to develop ideas that might well have led him to the kinetic theory of heat. As it was, and as I will now explain, his monograph was written with a totally wrong theory of heat in mind, and it did not matter. What was this wrong theory of heat?

---

[1] I spent the year 1959–1960 at the École Polytechnic which was still located in its ancient buildings in the Latin Quarter in Paris. The students were all in uniform. My patron, the late Louis Michel, a brilliant theoretical physicist, was a graduate. He too served for a time as a military engineer.

Carnot believed in the "caloric" theory of heat. He was in good company. For example, Benjamin Franklin measured heat flow down rods made of various materials to see how long it took to melt wax. He came to the conclusion that heat was a weightless fluid—caloric— which was conserved in all processes. He had correctly postulated that electricity flowed in currents. This time he missed. It was well known that friction produced heat. This was explained by arguing that when the objects rubbed against each other, caloric was transferred. There was, however, some evidence that argued against this. An American named William Thompson ended up in Bavaria where he was knighted. He became Count Rumford–Rumford being an old name for Concord, New Hampshire. While in Bavaria, Rumford occupied himself with, among other things, cannons. He was impressed by the amount of heat that is generated in cannon barrels by the explosion of gun powder and concluded that there could not be enough caloric stored to account for this. He made the same observation when he drilled out metal to make the barrels of cannons. Indeed, using the heat generated in such an operation he made water boil. He decided that something like a molecular theory of heat must be right and published this in 1798, to no effect. It took until the middle of the next century before the caloric theory of heat was finally disposed of. As it turned out, Carnot's belief in an incorrect theory of heat had no relevance to what he did.

Carnot was interested in what would constitute a perfect steam engine. This was a conceptual engine, but he thought that it might play some role in building a real steam engine. The key idea in Carnot's thinking about engines was the distinction he made between reversible and irreversible processes. In real life the processes that we encounter are irreversible. As the nursery rhyme has it, "All the king's horses/And all the king's men/Couldn't put Humpty together again." Reversible processes are, like Einstein's trains or elevators, thought experiments that illuminate laws of physics. In Carnot's "engine" there was a cylinder and a piston. The piston is assumed to work fictionlessly. In the cylinder was a gas—steam if you like–that was assumed to obey Boyle's law–a "perfect" gas. It turned out that Carnot's results did not depend on the specific choice of the working gas. In addition there were two heat reservoirs maintained at different temperatures. Heat–caloric in Carnot's case–can be added or subtracted from the reservoirs as needed. Two kinds of processes were allowed. In the first process heat is absorbed from the reservoir, which is at the same temperature as the ambient temperature of the gas, in such a way that the temperature of the gas remains the same while the pressure and volume adjust. This absorbed heat raises the piston. Such a process is called "isothermal" and, if carried out carefully, is reversible. The second kind of process is called "adiabatic." We now decouple the cylinder from the heat source and let the piston expand very slowly. This will reduce the temperature to that of the colder reservoir. If we do this, again very carefully, then it is reversible. Now we can put these two types of reversible processes together in a cycle which is a conceptual model for a reversible engine. It does not matter where we start the cycle or even in which direction we run it but conventionally it works like this (Figure 4.3).

Step 1. This is an isothermal expansion that maintains the temperature but during which the gas expands. Heat is absorbed.

Step 2. We now make an adiabatic expansion which reduces the temperature to that of the cold reservoir.

Step 3. We place the cylinder in contact with the cold reservoir and draw off the heat absorbed in Step 1.

Step 4. We make an adiabatic compression of the gas so as to return it to its initial condition.

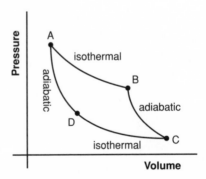

Figure 4.3. The four steps of Carnot's cycle.

These four steps are what is called the Carnot Cycle. Each of them is reversible. This makes it an efficient engine but what Carnot went on to show is that no engine can be more efficient than this.

To do this Carnot invoked a mantra that became a centerpiece for all the 19th-century development of thermodynamics; there is no such thing as perpetual motion. It is a curious statement if one thinks about it. How would you prove it? It is true that all the motions we have studied eventually come to a stop. But "eventually" is not "perpetual." Nonetheless, this doctrine was, and is, taken as a law of physics. What Carnot showed is that if there were a machine more efficient than his cycle it could be hooked to his cycle, run backward, and operate perpetually. Thus the Carnot cycle is the idealized limit of what can be achieved in making an efficient engine.

As I have mentioned, Carnot died young and his work was for some time not much appreciated. In midcentury he was rediscovered by a German physicist named Rudolph Clausius, who was born in what was then Prussia in 1822. It was in 1850, that Clausius wrote the paper that laid the foundations for thermodynamics. Clausius started from two principles. The first one was known as the mechanical equivalence of heat, something that we would call the conservation of energy. This had been established with some certainty by the experiments done by the British physicist James Joule—pronounced "jowell"—who was the son of a prosperous Manchester brewer. Joule had become convinced that the caloric theory was defective. It did not seem to account for the results of his experiments. In particular, he churned water with a paddle and found that this produced a measurable quantity of heat in an insulated can. This was difficult to account for by a transfer of caloric. In our terms, mechanical energy was being transformed into an equivalent amount of heat energy. This was one of Clausius's starting axioms.

His second principle was the proposition that you cannot transfer heat from a cold body to a hotter one without supplying some sort of energy. At first this seems crazy. How do you cool down soup? If one thinks about it, the air is cooler than the soup so we are transferring heat from a hot body to a cooler one, which is allowed. You can cool things in a refrigerator providing that you pay the electricity bill to run the motor. After he published his 1850 paper, Clausius put his second law in a different form. He introduced a concept that he named "entropy." He informs us that he took it from the Greek ητροπη–meaning a transformation or a turning. In classical thermodynamics the concept of entropy is somewhat slippery because what is well defined is the difference in entropy of two thermodynamic states–states that are characterized by quantities such as heat content and temperature.[2] In particular, if we have two identical heat reservoirs with heat content, Q, but at different temperatures, T and T^* then Clausius said that the difference in entropy between these states ΔS is given by

$$\Delta S = Q \left(\frac{1}{T} - \frac{1}{T^*} \right).$$

You cannot **prove** that this is the difference of entropy. It is how Clausius defined entropy. But how do you make use of this definition? What Clausius realized was that if you demanded that in all processes in a closed system entropy never **decreased** you could subsume in one law–this is the "second law of thermodynamics," the first being energy conservation–all the strictures on perpetual motion and the impossibility of transferring heat from a cold to a hotter reservoir without supplying energy from the outside and the rest.

To see how this works in a couple of examples let us begin with Carnot and his cycle. Carnot, as a consequence of his use of the caloric theory of heat in which heat is conserved in every transaction, would say

---

[2] In thermodynamics there is a third law that was formulated by the chemist Walther Nernst at the beginning of the 20th century. It says that as the temperature falls to absolute zero the entropy also falls to zero. Thus, given the third law, it is possible to define the absolute entropy of a system. The problem is that for, say, classical perfect gasses, the third law fails. For a classical perfect gas the entropy goes as the logarithm of the temperature and thus the third law is violated. Thus classically only the relative entropy is defined. I am grateful to Freeman Dyson for discussions of this.

that, at the end of his cycle, the reservoir would have exactly the heat content it started with. Furthermore, you have the same temperature initially and finally. Thus $\Delta S = 0$. In other words, if the cycle consists of reversible processes, then the change in entropy is zero. This is the limiting case of the second law. Now suppose that you try to make a transfer from a colder to a hotter reservoir without supplying additional heat energy. In this case $Q_f = Q_i = Q$, since no heat has been transferred. Thus $\Delta S = Q(\frac{1}{T_f} - \frac{1}{T_i})$. Therefore, if the final temperature is larger than the initial one, the change the change in entropy is negative which violates Clausius's second law. Therefore the second law prohibits such unassisted transfers. Clausius had his own mantra. It read *Die Energie der Welt ist constant. Die Entropie der Welt strebt einem Maximum zu.* "The energy of the universe is constant. The entropy of the universe tends to a maximum." By the time Clausius died in 1888, the first and second laws of thermodynamics were accepted parts of physics.

By this time, you may be saying, while this is all very nice, what does it have to do with the quantum? Patience.

Black bodies

On thermodynamic grounds Kirchhoff had concluded that the energy density and the spectral composition of radiation in a *Hohlraum*, surrounded by impenetrable walls of the temperature T, would be independent of the nature of the walls.

—Albert Einstein

We are now going to build a bridge between this 19th-century thermodynamics and the quantum. New actors will come on stage. The first of these is the German physicist Gustav Robert Kirchhoff who was born in 1824. He came from an socially conscious intellectual family who felt that being a university professor filled a civil obligation. Kirchhoff studied at the University of Königsberg and, while a student, made his first important discovery. His professor, Franz Neumann, was interested in electric circuits so Kirchhoff became interested in them too. He developed a set of relations that tell us how currents flow–"Kirchhoff's laws"–which we teach to undergraduates to this day. Kirchhoff first acquired a post in Berlin, then in 1850, in Breslau, where he met the chemist Robert Bunsen, who was somewhat older and was permanently attached to the University of Heidelberg. In 1854, Bunsen invited Kirchhoff to join him there. Bunsen was the inventor of the eponymous burner so beloved of chemists. He was in the process of using the burner to heat various elements to incandescence. Kirchhoff proposed that they use a spectrometer to measure the light spectrum, unique to each element, when it becomes incandescent. Over the next few years they discovered cesium and rubidium this way. They also used the spectrometer to extend work done earlier by Joseph Frauenhofer to study light coming from the Sun. Frauenhoffer had noticed that, at some wavelengths, instead of seeing bright lines, there were dark ones. Bunsen and Kirchhoff observed that if sodium vapor was put in front of the spectroscope these lines became even darker. Indeed, if they studied incandescent sodium through sodium vapor they saw the same effect. The conclusion they drew was that sodium could both emit and absorb light at the same wave length. They had no explanation for this, nor was any forthcoming until the advent of the quantum theory.

It was this work that led Kirchhoff to do the analysis of radiation that is most relevant to us. This analysis is philosophically somewhat akin to what Carnot did for engines. It does not deal with something that can be realized exactly in nature–although one must say that the background radiation left over from the Big Bang comes pretty close. It is an idealized situation that can teach us much. In his analysis he imagined a container whose walls can be heated up. It will turn out from Kirchhoff's analysis that it does not matter what the container is made of. Of course you do not want it to melt. Nor does its size and shape matter. I will now be a little anachronistic in describing what happens–a little, because by the end of the century this was the picture that had been adopted. Because they are heated, the electrons in the atoms of the material that composes the container walls will begin to oscillate. These oscillating electrons will radiate, and this radiation will enter the container, where some of it will be reabsorbed by the walls, and then reemitted. In due course, a situation of thermal equilibrium will be established in which the emission and absorption will just balance. At equilibrium the radiation will be characterized by some temperature T. When this happens there will be a distribution of different wave lengths of the radiation in the cavity. Some wave lengths will be favored and will be present with greater intensity than the less favored wave lengths. There is no reason to think that any wave length is excluded–from the longest to the shortest. It is just that they will have different intensities.

Now, we can imagine measuring these intensities and plotting the result in a graph. This graph will be representable by curve that, if we are lucky, will have a simple functional form. The question that Kirchhoff asked, and answered, is on what does this function depend? At first sight, we can think of a lot of things. It might depend on the size and shape of the cavity, or on what material it's made of. We know it will depend on the wave length, and surely it will depend on the temperature. As you heat things up, the dominant light you see changes color. That's what you mean, for example, when you say something is "white hot." What Kirchhoff showed is that, in fact, wave length and temperature are **all** this function depends on. Here is where the thermodynamics comes in.

Let us assume the contrary. Let us assume for the sake of argument that two containers made of different materials, with different sizes and shapes, have, at the same temperature different equilibrium radiation distributions. In one container some wave lengths may be favored and in the other, the same wave lengths will have, relatively speaking, less intensity. Now what we can do is to build a contraption that puts these containers together with a little door connecting them that we can open and close. With the door open, radiation will pass from one container to the other. Through the open door, radiation with the favored wave lengths will enter the second container to fill out its distribution. This is analogous to any kind of diffusion in which the diffusing material goes from places of greater to lesser density. But, in this case, it has done so between two regions at the same temperature with no external energy being involved. Now, if we close the little door, the two regions will again come to equilibrium, but at different temperatures, since the transport of the radiation has also transported energy. We have as a result a hotter and colder reservoir and can again open the little door. Heat will be transported back across from the hotter to the cooler reservoir. It is clear that we are starting to generate a perpetual motion machine, so the premise must be wrong. The two distributions must be the same. We can also draw a similar conclusion if we stick to one container and assume that the radiation distribution varies from place to place within it. The equilibrium distribution takes the same form no matter where we measure it in the container. Thus Kirchhoff drew the remarkable conclusion that this equilibrium radiation distribution was a universal function that depended only on wave length and temperature, and nothing else. But this was not all.

To appreciate the next step, let us examine the emission and absorption processes a little more closely. The emission will be governed by a relative probability function that tells us how likely the electrons in the wall of the container are to emit radiation with some given set of wave lengths. The walls of the container are at some temperature, T, which will be implicit in the discussion that follows. Let us call this function; "emission(λ)," where λ is the wave length. There will also be a corresponding absorption function. Here, there is a little subtlety. You can't absorb radiation that is not present. What is present is determined by Kirchhoff's universal function which I will call "universal(λ)." Thus the absorption is given by the product; absorption$(\lambda) \times$ universal(λ). At equilibrium, emission and absorption balance so that *emission*$(\lambda) =$ *absorption*$(\lambda) \times$ *universal*(λ) or $\frac{emission(\lambda)}{absorption(\lambda)} =$ *universal*(λ). Now comes the punch line. Kirchhoff imagined an object, which in 1862 he named a "black body," that is, a perfect absorber. It absorbs all wave lengths with equal avidity. In terms of our definitions *absorption*$(\lambda) = 1$. So that *emission*$(\lambda) =$ *universal*(λ). In nature there are no perfect black bodies, although I will shortly give you two very good approximations. What the equation says is that, for a black body which is in equilibrium with its radiation, we can measure *universal*(λ) by measuring the emission spectrum. This is something we might be able to get our hands on if we can identify something that acts at least approximately like a black body. We should be able to study the emission coming from it.

Here are two examples of entities that act like black bodies. The first is just a container with a small hole in it. Any radiation that falls on the hole will be absorbed by it and will have great difficulty escaping. Now, suppose you heat up the container. An equilibrium distribution will develop. A bit of this radiation will be emitted by the hole. By Kirchhoff's result, this radiation will be characterized by *universal*(λ). Thus, by measuring the radiation coming from the hole, we can measure the black-body spectrum of radiation. This, in essence, is how it was determined experimentally. The second example is much more exotic. It involves the early universe. Until something like 400,000 years after the Big Bang, matter in the early universe was in the form of a plasma. The electrically charged particles that constitute the plasma were primarily electrons and protons. There were also neutrinos and radiation left over from the original explosion. Radiation scatters from these electrons and an equilibrium is established. If you like, in this respect, the whole universe was a black body. Until the temperature fell to less than about 3,000°–which happens at something like 400,000 years–the hot radiation will not let neutral hydrogen atoms form. They would get ripped apart. But at a temperature of about 3,000°, the radiation is too cool to do this. Electrons and protons can now form electrically neutral hydrogen atoms and the radiation is free to expand with the universe. It has no charged particles to scatter from. Since nothing interferes, whatever black body distribution it had at the epoch of electron-proton combination, will be present now for us to observe, but, of course, at a vastly lower temperature. Indeed, this distribution has been observed with great precision. The black-body curve which arises from the experiments–I will show one later–is so good that, when you first see it, you think it must have been computer generated. It corresponds to a present temperature of about 2.73° above absolute zero, the average temperature of the present universe. With these examples in mind, we can now return to our historical narrative.

In 1879, the Austrian physicist Josef Stefan, who was Ludwig Boltzmann's teacher,[3] took an important step. We have not discussed the energy that a black body emits. To make the discussion not dependent on the size or shape of the hole in the container, one discusses the energy emitted per square centimeter of the surface of the hole, per second. One would guess that this "flux" would depend on the temperature of the container. The hotter it is, one would guess, the greater would be the energy flux. It is unlikely, however, that one would guess just how radical this dependence is. Stefan found that to fit the limited experimental data available, he had to assume that the flux varied as the **fourth** power of the temperature—T^4. If you, for example, double the temperature the flux will go up by factor of sixteen. This means that by doubling the temperature, a black body will radiate its energy away sixteen times as fast. In 1884, using methods of kinetic theory, Boltzmann was able to derive this result from first principles. But Stefan's empirical result was an important landmark.

The penultimate figure in this bridge building section is the German physicist Wilhelm Wien, who was born in East Prussia in 1864. When he was a young man he had difficulty deciding whether he wanted to be a physicist or a farmer. Einstein used to say that the best profession for a physicist would be as a lighthouse keeper, because you could just sit there all day thinking about physics. Of course Einstein became a university professor in some of the best universities in the world. Wien did not want to be a farmer to do physics. He wanted to be a farmer, period. By 1890, his parents were forced to sell the family farm, and Wien became a physicist for good. He took positions first at the University of Berlin, and ultimately at the University of Munich, where he remained until his death in 1928. From our point of view he did two essential things. One of them followed from a thermodynamic argument that goes beyond what we can present here, and the other from an inspired guess. Fortunately the consequences of these arguments are easy to state. Both have to do with

---

[3]My teacher, Philpp Frank, was Boltzmann's student. The chain gives me an odd feeling.

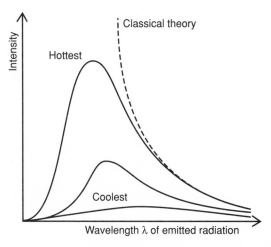

Figure 4.4. The short wave length end of the spectrum fits the Wien disiribution.

determining *universal*(λ). To appreciate the first one we have to recall that *universal*(λ) is also a function of the temperature. It is more correct to write is as *universal*(λ, T), a function of two variables. On its face this makes it much more difficult to find the function. The possibilities are now in two dimensions, which vastly widens the search. But, using thermodynamics, Wien proved a remarkable result. He showed that *universal*(λ, T) takes the form $\frac{1}{\lambda^5} \times$ *Universal*(λT). This means that the unknown function, which I have called "Universal" to distinguish it from "universal," is only a function of one variable, λT. This makes the problem of finding it even more tantalizing. In fact, using inspired guessing, Wien wrote down a form that turned out to be correct for part of the spectrum. Indeed, as we shall see, this was the very part that Einstein analyzed. I have a graph above that shows the full experimental spectrum at various temperatures. We will discuss the full graph soon. But, for the moment, I just want you to focus on the short wave length end (Figure 4.4). The long wave length end is labeled, for reasons I will also explain, "classical theory." At the short wave length end, you see from the graph that the spectrum falls rapidly to zero. This end of the spectrum is referred to as "ultraviolet" — beyond violet. Indeed, the fall at the ultraviolet end is exponential. This is precisely is what Wien proposed. His argument was pretty shaky, but his exponential form did seem to agree with whatever experimental data was available.[4] The year is now approaching 1890, and we have made some progress. But it is time introduce the last actor in this part of the drama, Max Planck.

↜ The Reluctant Radical

One should proceed as conservatively as possible in introducing the quantum of action into the theory, making only those changes in the existing theory that have proved to be absolutely necessary.
— Max Planck

In describing Planck the word "conservative" comes immediately to mind. It was almost genetic. Planck, who was born in 1858, in Kiel, where his father was a professor of civil law, came from a long line of lawyers and clergymen. He, and they, were deeply immersed in German culture. Planck was culturally, politically, and scientifically conservative. He and Einstein were antipodes. What they had in common was that they lived in troubled times. In World War I, a group of 93 very prominent German intellectuals, including Planck, signed a manifesto entitled *Appeal to the Cultured World*. It was a justification of German militarism. It indicted the French and British for "having allied themselves with Russians and Serbs, and presenting to the world the shameful spectacle of Mongols and Negros being driven against the white race." Needless to say, Einstein did not sign this document. Planck's oldest son Karl was killed in the war. After Hitler came to power, Planck, who had won the Nobel Prize in 1918, and was certainly one of the most recognized scientists in Germany, went to see him to try to convince him to temper his anti-Semitic laws against Jewish scientists in the universities. At the mention of Einstein,

---

[4]In formulae what Wien proposed was that $universal(\lambda) = a\frac{1}{\lambda^5}\exp(-\frac{b}{\lambda T})$ where a and b are constants that have to be adjusted to fit the data. He got this expression by using as an analogy Maxwell's formula for the way the energy is distributed among the molecules in a heated gas. It never occurred to Wien that if this analogy worked it might mean that black body radiation was behaving like a gas of particles.

Hitler went into a tantrum of rage. Planck retreated and spent the rest of the Nazi period attempting to salvage what he could of German science. In July of 1944, his younger son Erwin took part in an unsuccessful attempt to assassinate Hitler. He was arrested and died a terrible death at the hands of the Gestapo. Planck survived the war, dying in 1947.

As a child he was clearly gifted. He excelled in both music and mathematics. When he graduated from the *Gymnasium* in 1874, he briefly entertained the idea of becoming a pianist. However, he entered the University of Munich to study physics. He spent the year 1877–1878 in Berlin, where he attended lectures of Kirchhoff. Kirchhoff had moved to Berlin, and given up his experimental work, because an accident had left him crippled. He spent the later half of his life on crutches or in a wheel chair. But he was still lecturing. Planck was somewhat less than enthusiastic. He later wrote,

> I must acknowledge that I gained little from the lectures.... Therefore I could only still my need for continuing scientific education by reading works that interested me, and those naturally were ones relating to the energy principle [the conservation of energy]. In this way I came upon the papers of Rudolph Clausius, whose clarity of expression and thought made a powerful impression. With growing enthusiasm I worked my way deeply into them. What I particularly admired was the exact formulation of the two laws of thermodynamics....

This is, of course a reference to the first law—the conservation of energy—and the second law—that entropy never decreases. As Planck then viewed these two laws they were on the same footing. Both were absolute statements about nature, which appealed to Planck's scientific conservatism. Planck was also a Machian anti-atomist. He soon found himself in a conflict with Boltzmann.

Boltzmann was a confirmed atomist who believed that the laws of thermodynamics reflected the statistical behavior of molecules. Heat, for example, was simply a manifestation of the disordered motion of molecules. But, when it came to entropy there was a dilemma. It had been posed to Boltzmann by his senior colleague and teacher Loschmidt.

Molecular collisions were reversible. To any collision there was a re-verse collision with the initial and final momenta interchanged. If you filmed the collisions you could not tell whether the film was running forward or backward. How then did what appeared to be an absolute law of irreversibility–that entropy never decreased–arise? Boltzmann's solution was to argue that, unlike the first law which says that energy is **always** conserved, the second law says only that entropy **probably** never decreases. It is astronomically unlikely that "humpty" will spontaneously be put together again, but it is not impossible. Boltzmann spent a number of years showing explicitly how systems evolve towards more and more probable configurations, which is what he meant by an increase in en-tropy, always allowing for the possibility that there can be fluctuations. This is what, according to Boltzmann, the second law means.[5] This in-terpretation was, in his early years, an anathema, to Planck. In 1882, he wrote a paper in which he concluded,

> Consistently developed, the second law of the mechanical theory
> of heat is incompatible with the assumption of finite atoms. It can
> therefore be foreseen that the further development of the theory
> will lead to a battle of these two hypotheses in which one of them
> will perish. An attempt to predict the conflict's outcome with pre-
> cision at this time would be premature. Nevertheless, a variety of
> present signs seems to me to indicate that atomic theory, despite
> its great success, will ultimately have to be abandoned in favor of
> continuous matter.

By the turn of the century Planck had changed his mind.

It was almost inevitable that Planck would become involved with black-body radiation. In 1889, after Kirchhoff's death, he succeeded him in Berlin. Berlin was the world center for the study of black-body ra-diation. Wien was there when, in 1896, he proposed his form of the

---

[5]Boltzmann expressed this in a formula If, S, is the entropy of a state, and, P, is the probability of that state's occurrence then $S = - \mathrm{k} \log(P)$ where k is a constant–Boltzmann's constant. Boltzmann was so pleased with this expression that he had a version of it engraved on his tombstone.

distribution. In addition, there were two very powerful experimental teams measuring the spectrum of the radiation, that of Otto Lummer and Ernst Pringheim and that of Heinrich Rubens and Fernand Kurlbaum. What experimental evidence there was when Planck began considering the problem, agreed with the Wien law. For Planck, the problem of black-body radiation was therefore how to derive the Wien law from first principles. For the next decade he produced various arguments each one of which he was sure was definitive, and each one of which was flawed. Whatever else, it gave him a mastery of the techniques. It also gave him a constant. This is one of the strangest aspects of the story. The Wien formula had two arbitrary constants that could be adjusted to fit the data. One of them had the peculiar dimensions of energy × time. Planck argued that it must be a fundamental constant of nature since the black-body spectrum was universal. He realized that this constant enabled him to complete a set of units that had universal meaning. What I mean is this. Take, for example, the usual unit of length, the meter. It was originally defined as a ten millionth of the length of the meridian through Paris from a pole to the equator. How would you explain this to an extraterrestrial? But you could explain the charge of the electron, or the strength of the gravitational force. These are universal units. There are no references to Paris. You could also explain Planck's new constant, assuming the extraterrestrial could construct a black body. As Planck wrote, his units would be "independent of particular bodies or substances, would necessarily retain their significance for all times and all cultures, including extraterrestrial and non-human ones." Planck, with his conservative bent, and love of the universal, was enthralled by this new set of units. We are enthralled by them too. To take an example, if we call Planck's constant h, The gravitational constant, G, and the speed of light c, then the Planck length l_p is given by $l_p = \sqrt{\frac{Gh}{c^3}} \simeq 4 \times 10^{-33}$cm. As we will now see, it was his new constant that heralded the onset of the quantum. What is strange about this was that he discovered this unit before he discovered that there was a quantum.

By the fall of 1900, it was becoming clear that the wheels were coming off. Both of the experimental groups had been able to extend their measurements to longer wave lengths and these results did not agree with Wien. Planck apparently learned about the new results on the afternoon of October 7, when Rubens and his wife came for a visit. For long wavelengths the spectral function appeared to be going to zero proportional to the wave length, i.e., $universal(\lambda T) \approx \lambda T$. Thus at short wave lengths the function went to zero exponentially, while at long wave lengths it went to zero with λ. Planck's problem was to find a distribution that did both of these things. By that evening, Planck had come up with one, largely inspired guess work, based on years of immersion in the problem.[6] The new distribution had two virtues. It agreed with Wien for small wave lengths and with the new results at longer wave lengths. Indeed, it agreed with experiment at all the wave lengths that had been measured. It still does. Figure 4.4 showed Planck's distribution for three temperatures. The graphs were prepared using Planck's formula–and not the experiments. But Figure 4.5 is the measured black-body spectrum left over from the radiation produced by the Big Bang, with the present temperature of about 2.73 degrees above absolute zero. You cannot distinguish the experimental from the theoretical curve.

Now Planck had a new problem; how to derive his distribution.

In accounting for what Planck did in 1900, and 1901, there is some difficulty. Part of the problem is that Planck's papers, taken on their own terms, are opaque. But the real problem is that Planck was trying to do something that was impossible. He was trying to derive his distribution from classical physics. He spent a decade trying. Late in life he wrote a scientific autobiography in which he says,

> My futile attempts to fit the elementary quantum of action into the classical theory continued for a number of years, and they cost me a great deal of effort. Many of my colleagues saw in this something bordering on tragedy. But I feel differently about it.

---

[6]In the notation of footnote 4 what Planck did was to replace $e - \frac{b}{\lambda T}$ by $\frac{1}{e^{\frac{b}{\lambda T}} - 1}$. As my teacher Philipp Frank used to say, those of you who know a little of mathematics can show that this distribution morphs into that of Wien when the wave length becomes small.

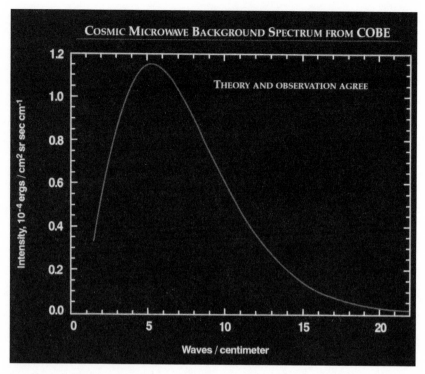

COSMIC MICROWAVE BACKGROUND SPECTRUM FROM COBE

THEORY AND OBSERVATION AGREE

Intensity, 10⁻⁴ ergs / cm² sr sec cm⁻¹

Waves / centimeter

Figure 4.5. The experimental black-body spectrum, for the cosmic microwaves cannot be distinguished from the theoretical.

> For the thorough enlightenment I thus received was all the more valuable. I now knew for a fact that the elementary quantum of action played a far more significant part in physics than I had originally been inclined to suspect.

What there is no disagreement about was how he went about the problem. It was an extension of what he had done to try to derive the Wien distribution.

In the first place, there was the model. The electrons in the atoms that make up the walls of the black body container are set into oscillation when the walls are heated. These oscillating electrons–which Planck referred to as "resonators"–emit–and absorb–radiation. An equilibrium will be reached in which emission and absorption balance. At this equilibrium the oscillators will have an entropy. Planck showed that if he knew the functional form of the entropy he could use the laws of thermodynamics to work his way back to the distribution of radiation. This is one of the things he came to understand in the decade he worked on this problem before 1900. Thus it came down to finding this entropy. By 1900, Planck had accepted Boltzmann's idea that to find the entropy you had to find the probability of various configurations of the distribution of energy among the oscillators. As the system evolves towards equilibrium the system evolves towards more probable states until it reaches the state of maximum probability at equilibrium. So the problem came down to finding the most probable distribution of the energies among the oscillators. This would be the black body distribution. It was precisely at this point that the quantum came in.

Finding probabilities is a counting problem. You count the number of faces on a die — six — and you conclude that the probability of throwing a one is one-sixth. Planck was forced to reduce his probability question to a counting problem. It is here that the historians disagree as to how he went about it. I will tell you the story that I first learned from my teacher Philipp Frank. It is the one that we tell our students. Put anachronistically, according to this version, Planck "quantized" the oscillators. Here is what this means. A classical oscillator can absorb and emit radiation of any energy from zero to infinity. A quantized oscillator has restrictions. Suppose we call the basic energy unit E, then the quantized oscillator can only emit and absorb radiation that has the values E, 2E, 3E, 4E, and so on. These units of energy Planck called "quanta." Professor Frank had a homely analogy. He said that it is like buying and selling beer in pints and quarts. I suppose the classical oscillator was like buying and selling beer on tap, although I do not recall Professor Frank as having made that analogy. In terms of these quantized energies Planck could carry out his counting to compute the probabilities. There was some hanky-panky in his counting procedure as well. The full meaning of this was not understood for many years.[7] Be that as it may, with these rather

---

[7] Not to be too laconic about this, when Boltzmann did his counting he assumed that his molecules were "distinguishable." Metaphorically speaking they came

odd assumptions Planck was able to derive his distribution provided that he set $E = h\nu$. Here, ν is the frequency of the radiation and h is the constant with the strange dimensions that Planck had introduced in his version of the Wien distribution. Using his 1901 data, Planck found $h = 6.55 \times 10^{-27}$ erg. seconds, while the modern value is 6.63×10^{-27} erg. seconds. The "erg" being a standard unit of energy. Recall that there was a second constant to be determined in the Wien distribution. This one can be related to Loschmidt's number; the number of molecules in a cubic centimeter of a standard gas. From the data Planck found 2.76×10^{19} molecules per cubic centimeter to be compared to the modern value of 2.69×10^{19} molecules per cubic centimeter. It was at the time the most accurate determination that had been made.

I have given you the standard version of how Planck made his discovery, but historians who have studied his papers and correspondence carefully do not think that it is correct. They may well be right. They point out that nowhere in his early papers does Planck say anything about quantizing individual oscillators. What he does do is to take a group of oscillators and stipulate that their total energy is an integer multiple of a basic quantum unit. This was a device that Boltzmann used when he computed probabilities. But then, at the end of his calculations, Boltzmann allowed the energies to become continuous again. Planck could not do this because it gave him the wrong answer. If he did it he was lead back to the Wien distribution. Planck was, at least temporarily he thought, stuck with the quantum. Because of the way he did it, he did not think at first that he had done anything radical. He was sure that if he kept working at it he would find a way of deriving his distribution from classical physics. The real revolutionary is now coming on stage.

---

in different colors. Planck, with no justification, assumed that his quanta were "indistinguishable." This completely changes the counting problem. If he had not made this assumption he would have ended up with the wrong distribution. Nearly three decades later this was understood after the creation of the quantum theory.

∽ PHOTONS

> It appears to me, in fact, that the observations on "black-body ra-
> diation," photoluminescence, the generating of cathode rays with
> ultraviolet radiation, and other groups of phenomena related to the
> generation and transformation of light can be understood better on
> the assumption that the energy in light is distributed discontinuously
> in space."
>
> —Albert Einstein

Of Einstein's four "Miracle Year" papers, his paper on black-body radiation is to me the most miraculous. This is to take nothing away from the other papers, but here he is truly groping in the dark. Here he is at age twenty-six, with no academic job, barely known in the physics community, and no other professional physicists to talk to, not even an adequate library, about to take down the entire edifice of classical physics. I have often thought that if Einstein and Maxwell had been able to sit down and have a chat, Maxwell would have understood relativity in an hour. With Newton you would have had to explain too much. The Brownian movement paper was certainly a fine paper but, after all, it was done almost as well by someone else—Smoluchowski. No one else could have written the black-body paper and no one else believed the results. When Planck, and three colleagues, proposed Einstein for membership in the Prussian Academy they wrote the following:

> In sum, one can say that there is hardly one among the great prob-
> lems in which modern physics is so rich to which Einstein has
> not made a remarkable contribution. That he may sometimes have
> missed the target in his speculations, as, for example, in his hypoth-
> esis of light-quanta cannot really be held too much against him, for
> it is not possible to introduce really new ideas even in the most
> exact sciences without sometimes taking a risk.

This was written in 1913! What enabled Einstein to get through this maze was his incredible intuition. He seemed to have some kind of internal guidance system that told him what was correct–a pipe line to the secrets of the Old One, Einstein's affectionate way of referring to God. You cannot be taught this. You have it or you do not, and the great physicists have it–no one more than Einstein in these early years. The title of the paper is interesting. In English translation, it is "Concerning a Hueristic Point of View about the Creation and Transformation of Light. "Heuristic"– "*heuristisch*" in German– is such a curious choice of word. I have never seen it in the title of any other physics paper. It seems to come from the Greek. The dictionary definition is "Of or relating to a general formulation that serves to guide investigation."[8] Unlike relativity, he is not offering here a full-blown theory. Part of his genius was to understand that, at the time he wrote the paper, there was not, and could not be, such a theory. He was proposing some guide as to what such a theory had to explain.

The paper starts out with an introduction in which Einstein briefly summarizes the successes of the classical theory of electromagnetism– Maxwell's theory.

This theory works with continuous distributions of electromagnetic energy. It accounts for a vast domain of electromagnetic phenomena. It is the theory that Einstein discusses in his relativity paper. In that paper, written afterwards, there is only one hint that for some purposes the theory has to be modified to allow for finite numbers of "energy quanta" instead of the continuous electric and magnetic fields. In the eighth section of the relativity paper he computes the energy and frequency of what he calls a "light complex." He notes laconically that "It is noteworthy that the energy and frequency of a light complex vary with the observer's state of motion according to the same law." One waits in vain for the other shoe to drop. Why does he not simply say that $E = h\nu$, something that he had proposed a few months earlier? In this earlier paper he puts it very starkly,

> According to the presently proposed assumption the energy in a
> beam of light emanating from a point source is not distributed

---

[8]This definition, and much more about the word, can be found at http://www. websters-onlinedictionary.org/definition/english/he/heuristic.html.

continuously over larger and larger volumes of space but consists of a finite number of energy quanta, localized at points in space, which move without subdividing and which are absorbed and emitted only as units.

In terms of Professor Frank's beer analogy, not only is beer bought and sold in pints and quarts, but even in a barrel you would find it localized in pints and quarts. No draft beer. It is clear how radical this is. Nothing had prepared physicists for this. The rest of the paper is an examination of "the proposed assumption."

Before doing this, Einstein has a section of the paper which he calls "Concerning Certain Difficulties in the Theory of Black-Body Radiation." In this section he answers the question that Planck should have answered. Suppose you do not know the "right answer"–Planck's distribution–but simply apply classical physics consistently to the problem, what distribution of the radiation do you find? Einstein does this and finds that the resulting distribution is neither Planck's nor Wein's. It is plotted in Figure 4.4 under the label "Classical theory." It is just the long wave length distribution that Planck had used after he had spoken with Rubens. This classical distribution which is simply proportional to the wave length cannot be right at all wave lengths. Not only does it not fit the data, but it leads to an absurdity. You can use it to compute how much radiation energy is contained in any cubic centimeter in the container. What you find is that the amount is **infinite**! This absurdity came to be called the "ultra-violet catastrophe" in honor of that fact that as λ gets smaller—goes towards the ultra-violet—the contribution to the energy at these wave lengths goes off to infinity. Classical physics, which is what is assumed to derive this distribution, has broken down.

Einstein was not the only one to have reached this conclusion. In 1900, John Strutt, better known a Lord Rayleigh, published a paper in the British journal *Philosophical Magazine* in which he presented his own radiation formula. It was an odd amalgam of the classical result and Wien formula. He offered no real explanation of how he had arrived at this formula. If Planck was aware of it he did not mention it in his papers. Einstein certainly was not aware of it. Nor was he aware of

Rayleigh's 1905 paper in which he spells out in detail his derivation of the classical distribution. This derivation is part of what we teach our students. But we have to amend it to take account of the quantum effects. Rayleigh did not bother with Planck's radiation oscillators. He went right to the radiation in the cavity. He wanted to count how many different wavelengths of the radiation could be fitted into the cavity. The restriction is that each wave has to fit inside the cavity with nothing slopping over. The answer he got was off by a numerical factor that was later discovered by James Jeans, the British astrophysicist. The resulting distribution is usually referred to as the Rayleigh-Jeans distribution. I think it might well be called the Rayleigh-Jeans-Einstein distribution. To make the final step to get it, Rayleigh used a fact from classical physics that each of these radiation "modes" in equilibrium has the same average energy. The energy is equally partitioned among these modes. This law was embedded in classical physics and led at once to the classical distribution and all the attendant difficulties.[9] Rayleigh and Jeans both understood this but were unclear what to do about it. I do not think they saw in it the end of classical physics. In the new physics the energy is not equally distributed among the modes and this is what leads to the Planck distribution which Einstein clearly understood. He discusses the meaning of Planck's derivation in the third section of the paper.

To me, the miraculous section of this paper is the fourth which has the ponderous title, "Limiting Law of the Entropy of Monochromatic Radiation for Small Radiation Density." You can as much guess what

---

[9]Not to be overly mysterious about this, there was in classical physics a principle that was called the "equipartition of energy." In this instance what it implied was that each of these modes had equal average energy and that this energy was given by kT. Here T is temperature and k is the Boltzmann constant which has a value of approximately of 1.38×10^{-16} ergs per degree. The important point is that this energy did not depend on the frequency of the radiation, which is what lead to the disaster. In the Planck case the energy does depend on the frequency and is given by $\frac{h\nu}{e^{\frac{h\nu}{kT}}-1}$. This expression has the property that for low frequencies it goes over to the classical answer and for high frequencies it is cut off by the exponential so that the catastrophe is avoided.

this section is about from this than you can tell from the title of the painting "La Giocanda" that you are about to see the *Mona Lisa*. As I have discussed, Planck's tactic for attempting to derive distributions was first to derive an expression for the entropy from which he could derive the distribution. Einstein turned this around. He started from the observed distribution, in particular from the Wien end where the new physics was to be found, and from it calculated the entropy. Planck had also found this expression but he had not understood what it meant. Einstein shows that it is the same expression you would find if you were considering a dilute gas of particles. The difference is that these "particles" have energies that are proportional to their frequencies; i.e., $E = h\nu$. They, are in this respect, not like the molecules of, say, a volume of hydrogen gas where the energy depends on the speeds of the particles. These "particles" all move with the speed of light. Recall from the relativity chapter, this means they are massless. Where does this leave us? The classical part of the black-body distribution is derived from the assumption that the radiation in the black body has a wave-like character. But the Wien part of the distribution is derived from the proposition that the radiation has a particle character. And in the middle? Here you have the first instance of the wave-particle duality of light. It would haunt Einstein for the rest of his life. He never came to terms with it.

In the next two sections of the paper Einstein sharpens his particle analogy and then come the last three sections of the paper in which Einstein discusses tests for this new idea. The most famous section of the entire paper is the eighth. It has the cumbersome title "On the Production of Cathode Rays by Irradiation of Solid Bodies." To explain it a little history is in order. In 1886, the German physicist Heinrich Hertz performed the experiments that successfully showed that Maxwell's prediction that light was an electro-magnetic wave was right. In the course of this he found that a spark he was using to detect the radiation was enhanced if ultra-violet light struck the metal of his detector. He had no explanation for this but, being a good experimenter, noted it. As it turned out, in the same set of experiments he had, unknown to him, detected both the particle and wave properties of light. But that realization would only come later. In 1897, the British physicist J.J. Thomson identified the electron as the particle that was emitted from the cathode–the negatively charged component of a vacuum tube. These electrons became known as "cathode rays." In 1899, he showed that these same cathode rays were emitted when ultraviolet light was shone on a metal. What Hertz had actually seen were these electrons. The next important step was taken just after the turn of the century by the German physicist Philipp Lenard. Lenard had a carbon arc light whose intensity he could vary by a factor of a thousand. This light was shone on a metal plate and the emitted electrons collected on a detector plate. He was able to measure the energy of the electrons. The result was totally unexpected. Common sense, and indeed classical physics, would have predicted that when the intensity of the light was increased the emitted electrons would have a higher energy. But this is not what happened. In general, the electrons are emitted with various energies but one can focus on the maximum energy. What Lenard found was that this energy did not change when the intensity was increased. What happened was that number of emitted electrons carrying this energy increased. He then very cleverly broke up the light from his carbon arc into various frequencies using a spectrometer. He seemed to find that the higher frequency components of the light caused electrons of higher energies to be emitted, but the data was not conclusive.

In this section of Einstein's paper he explains everything in one equation that is so simple that you could teach it to high school students. It was this equation for which Einstein was awarded the Nobel Prize for 1921, which he collected in 1922. The citation of the Royal Swedish Academy that awards these prizes is quite marvelous. It reads,

ROYAL SWEDISH ACADEMY has at the assembly held on November 9,1922, in accordance with the stipulation in the will and testament of Alfred Nobel dated November 27, 1895, decided to/independent of the value that/(after eventual confirmation) may be credited to the relativity and gravitation theory/bestow the prize/that for 1921 is awarded to the person within the field of physics who has made the most important discovery or invention/

to Albert Einstein being most highly deserving in the field of theoretical physics/particularly his discovery of the law pertaining to the photoelectric effect.[10]

The "photoelectric" effect is the shorthand name for the phenomenon of the emission of electrons by light. Incidentally, the Prize was worth about 32,000, 1922 dollars, all of which went to his ex-wife as part of a divorce settlement that had been made several years earlier. What then is the equation? I will not use the notation of Einstein's paper since he has a way of rendering Planck's constant "h," which is a consequence of how he defines the Wien distribution, that is rather ungainly. In modern notation the equation is simply

$$E_{\max} = h\nu - P.$$

Here $E_{\max}$ is the maximum energy an electron can have when it is liberated by a quantum that has energy $h\nu$. The reason that it cannot have all the energy is that some is needed to allow the electron the escape from its binding to the metal surface on which the light is incident. This is what "P" stands for. This equation is simply the conservation of energy. It says that if N quanta of energy $h\nu$ are incident on the surface then N electrons with a maximum energy of $E_{\max}$ can be released. It furthermore says that the energy of the electrons increases with the frequency of the light. This explains Lenard's results. The first sentence of this section of Einstein's paper reads, "The traditional view that the energy of light is distributed continuously through the region illuminated by the light runs into great difficulty in trying to explain photoelectric phenomena, as was outlined in a trail-blazing paper by Lenard." Two decades later Lenard had become a rabid anti-Semitic Nazi and all of this was, according to him, "decadent Jewish physics."

In the next, and final section, "Afterward," I will describe the experiments that confirmed this equation and end with a brief overview of Einstein's attitude towards the quantum theory as the theory evolved.

---

[10] I am grateful to Gerald Holton for supplying the translation of the Swedish original.

⤳ AFTERWARD

> I cannot make a case for my attitude in physics which you would consider at all reasonable. I admit, of course, that there is a considerable amount of validity in the statistical approach which you were first to recognize clearly as necessary given the framework of the existing formalism. I cannot seriously believe in it because the theory cannot be reconciled with the idea that physics should represent a reality in time and space, free from spooky actions at a distance.
> — Einstein to Max Born, 3 March 1947

The definitive experiments on the photoelectric effect were carried out by the American physicist Robert Millikan in the years from 1914 to 1916. They confirmed Einstein's equation and for them Millikan won the Nobel Prize in 1923, two years after Einstein. One might think that Millikan's experiments would have settled the matter in the sense that the quantum would have achieved universal acceptance. This was far from the case and Millikan is a prime example. In 1917, Millikan published a book called *The Electron*, in which he describes these experiments. He writes,

> Despite then the apparently complete success of the Einstein equation, the physical theory of which it was designed to be the symbolic expression, is found so untenable that Einstein himself, I believe, no longer holds to it, and we are in the position of having built a very perfect structure and then knocked out entirely the underpinning without causing the building to fall. It stands complete and apparently well tested but without any visible means of support, and the most fascinating problem of modern physics is to find them. Experiment has outrun theory, or, better, guided by erroneous theory, it has discovered relationships which seem to be of the greatest interest and importance, but the reasons for them are as yet not at all understood.

Millikan was correct in the sense that the quantum theory, which was barely in its infancy, was not well understood. But his notion that

Einstein had abandoned the quantum as an explanation of the photo-electric effect is absurd. Indeed, typically, after having published his explanation of the photoelectric effect Einstein found another application of the quantum idea. This had to do with applying Planck's oscillator model with individually quantized oscillators to the study of the absorption of heat in solids. The history of this subject goes back to 1819, when two young French physicists, Pierre Louis Dulong and Alexis Thérèse Petit studied the heat absorption of a variety of materials — mostly metals. They found that, taking a standard amount of the materiel, the amount of heat needed to raise the temperature by, say, one degree, appeared to be the same for all these elements and did not depend on the temperature of the material. This law was readily derived from a model involving classical oscillators. The problem was that by the turn of century it seemed to be breaking down at the lower temperatures. In 1906, Einstein supposed that in reality the energies of the oscillators were "quantized." Only integer multiples of the basic energy unit was allowed. Using this assumption, he found a new law for the specific heat which agreed with the old one at the higher temperatures and with the experimental data at the lower ones. This study was the first to apply the quantum theory to solids. It persuaded some people who had not taken the quantum seriously before to take it seriously now. The next great step was taken in 1913, by Niels Bohr. What Bohr did was to quantize the electron orbits in atoms. The electrons were only allowed to travel in restricted orbits. When an electron "jumped" from a higher to a lower orbit it emitted a quantum of radiation equal to the energy difference between the two orbits. This incidentally explained the results of Bunsen and Kirchhoff on the dark sodium spectral lines. Absorption of light was subject to the same quantum rules. For light to be absorbed it had to have the right energy to induce of these quantum jump. Sodium light could be absorbed by sodium vapor because it had the right energy. Einstein was extremely enthusiastic about Bohr's work.

In 1916, Einstein presented a new way of looking at Planck's distribution. It cannot exactly be described as a derivation because there was as yet no theory from which to derive it. But one of the basic ideas became decades later the basis of the laser. The idea was to consider a model in which, say, an electron, could be in one of two energy states. There was a state of least energy, the so-called "ground state" and an excited state of greater energy. This system was assumed to be in a bath of radiation at some temperature. Two obvious things can happen. If an electron is in the ground state it can absorb a radiation quantum and jump to the excited state. Once in the excited state it can spontaneously emit a quantum of radiation and jump back to the ground state. But Einstein discovered that there was a third process. The radiation bath can stimulate the excited electron to emit a quantum. It was this stimulated emission that was exploited in the laser since it led to a method of amplifying the intensity of the emitted radiation.

In 1923, the French theoretical physicist Louis de Broglie in a proposed PhD thesis, suggested that particles like electrons might also have a wave character. This completed the wave-particle duality. Light was both a wave and a particle, and an electron was both a particle and a wave. De Broglie's thesis advisor, Paul Langevin, sent a copy of the thesis to Einstein, who replied that he found the ideas interesting. The wave nature of the electron was demonstrated experimentally four years later. At first, it was assumed that these were waves like light waves in that they oscillated in ordinary space. They might act as "guides" for the particles. Indeed, in 1926, the Austrian physicist Erwin Schrödinger found the equation that bears his name which described these waves. Einstein was delighted with this until Max Born, and others, showed that these waves were different. In fact they were waves of probability. Where they had large amplitudes, a particle was most likely to be found. It was at this point that Einstein and the quantum theory parted company. Indeed, in 1926, in a famous and often quoted letter to Born he wrote,

> Quantum mechanics is certainly imposing. But an inner voice tells me that it is not yet the real thing. The theory says a lot, but it does not really bring us any closer to the secrets of the "Old One." I, at any rate, am convinced that *He* is not playing dice.

Einstein then entered a period in which he tried to show that the theory was wrong. This inspired monumental debates with Bohr which, in truth, Einstein lost, but we all won since they clarified the theory. Finally, Einstein decided that the theory was not a complete description of "reality." Usually when physicists begin talking about reality it is a bad sign. He tried for the rest of his life to produce a theory that would replace quantum theory. But he was doing this without any guidance from experiment so there was little chance for success. Many people tried to induce him back into the mainstream. John Wheeler told me that when Feynman, who was his student, produced a new formalism for the quantum theory Wheeler thought that it was so beautiful that Einstein would surely be converted. Of course he was not. I think that Einstein was in his heart a classical physicist. Poets, as Cocteau once said, tend to sing from their family trees. Einstein's family tree was classical physics—the physics he learned as a student–and while quantum theory was, in a sense, his child, it was not a child that he was able to adopt.

Our presentation of the four essays is now done. In the brief epilogue I will trace the arc of the remainder of Einstein's life.

Epilogue: Afterword

When he spoke in Zurich in 1909 for the first time on special relativity, it was neither at a university nor at the ETH, but in a room of the Carpenter's Union at a town restaurant. For writing he had only a small blackboard on which he drew a horizontal line; it was a space of one dimension that he was going to relate to his new notion of time. He began by saying, "Consider at each point of this straight line a clock-that is, infinitely many clocks." After having developed his theory for more than an hour, he suddenly stopped and excused himself for having spoken for so long. "How late is it, because I have no clock?

—Louis Kollros a classmate of Einstein at the ETH

The theory of relativity to which Minkowski later gave an appropriate mathematical form attracted the attention of many people to the

new mechanics. The overthrow of the fundamental conceptions of kinematics surprised those who had not followed the historical evolution of these problems which we have just sketched. The apparent generality of the solution of the problem of space and time met the desire of that period of unifying and synthesizing science. That is why the theory of relativity excited the young devoted to the study of mathematical physics; under the influence of that theory they filled the halls and corridors of universities. On the other hand, the physicists of the former generation, whose philosophy was formed under the influence of Mach and Kirchoff, remained for the most part skeptical of the audacious inventors who allowed themselves to rely upon a small number of experiments, still debated by specialists, to overthrow the fundamental tests of every physical measurement.
— Max Abraham, a theoretical physicist in 1914

One of the more striking things about the young Einstein is that he had an absolutely clear understanding of the value of what he had done. His communication to Conrad Habicht, which I quoted in the last chapter makes this evident. He was also ambitious. He expected that there would be considerable reaction to his four papers from other physicists, and there was almost none. The only manifestation of interest that was apparent to Einstein was the visit to Bern by the equally young Max von Laue. There was some interest–Max Planck was an example–but this was not directly communicated to Einstein. This had to do with relativity. As far as the quantum paper was concerned, there was almost no interest. This may seem odd to us. We understand that the quantum paper heralded the end of classical physics. Einstein certainly understood this. But this is not how it appeared to the physics community. The historian of science Thomas Kuhn made a study of the number of papers published on black body radiation and the quantum from 1905 to 1914 Until 1910, there were less than ten authors per year. After 1910, the numbers begin to exponentiate. There was nonetheless among the physicists of Einstein's generation, a reluctance to accept the radical consequences of the quantum theory. Einstein himself reflected this. His last positive contribution to the theory was inspired by a letter he received in 1924, from a young Indian physicist named Satynendra Nath

Bose.[1] Enclosed with the letter was a paper that Bose wanted Einstein's help in publishing in one of the German language physics journals. The paper was written in English and Einstein thought that it was important enough to translate himself into German. What Bose had done was to derive the Planck distribution by making explicit the counting procedures that Planck had used implicitly. But Einstein realized that there was a consequence of these procedures that had gone unnoticed. There was a class of particles-now known as Bose-Einstein particles — to which the same counting procedures had to be applied. For example, the nucleus of the common isotope of helium is such a particle. What Einstein noticed was that if one lowered the temperature of a collection of these particles there was a critical temperature at which they would condense into one quantum state. This remarkable phenomenon — which is known as Bose-Einstein condensation, although Bose had nothing to do with it — is now a very active branch of theoretical and experimental physics.

The reception of relativity was also mixed. There were many physicists such as W.F. Magie, whom I quoted before, who did not have any understanding of the theory and were ineducable. But there were also several physicists who actually worked in the theory, but who saw it as simply an adumbration of what Lorentz had done. These people did not grasp what Einstein referred to as the "kinematical" part of his paper. They wanted to proceed from specific electron models of matter to derive such things as the Lorentz-Fitzgerald contraction. They thought that what Einstein had done–to start with two axioms, relativity and constancy–and then to show how an analysis of space and time led to these results, was almost a form of cheating. Where were the calculations? Lorentz himself seemed to have mixed feelings about what Einstein did. A perfect example is a statement, one of his last on this subject, which he made in the course of some lectures he gave at the California Institute of Technology in 1922. He noted that, "As to the ether... though the conception of it has certain advantages, it must be

---

[1] People who know Bengali have informed me that the accepted pronunciation of "Bose" is "Bosh."

admitted that if Einstein had maintained it he certainly would not have given us his theory, so we are grateful to him for not having gone along old-fashioned roads." What possible "advantages" did the conception of the aether have in 1922? Established ideas die hard.

Poincaré is another curious example. Of Einstein's predecessors he was the only one who understood the issues. One would have thought that he would have embraced the theory. In fact, he made some important mathematical contributions to it. But in his papers he never once mentioned Einstein. He too seemed to think that he was adumbrating Lorentz. He was certainly aware of Einstein. Indeed, in 1911, he wrote a very enthusiastic letter of recommendation to the ETH in Zurich which was considering Einstein for a position. He wrote,

> Einstein is one of the most original minds that I have known; despite his youth he has already achieved a very honorable rank among the foremost scholars of our time. What we can, above all, admire in him is the facility with which he adapts himself to new concepts and draws all the consequences from them. He does not remain attached to classical principles and in the presence of a physics problem is prompt to realize all of its possibilities. This translates itself immediately in his mind by the prediction of new phenomena, which can be verified by experiments. I do not mean that all of his predictions will be confirmed when they are eventually tested. Since he searches in all possible directions, one should, on the contrary, expect that most of the paths that he follows will lead to an impasse; but one may also hope that one of the directions that he has pointed to will be the true one; and that is enough. This is the way that one ought to proceed. The role of mathematical physics is to ask questions; it is only experience that can answer them. The future will show, more and more, the worth of Einstein, and the university which is able to capture this young master is certain to gain much honor from this operation.

It is interesting that this very generous recommendation was written the year that the two men encountered each other for the first time at the Solvay Congress. Recall that Einstein had gotten a rather negative impression of Poincaré at this conference. The late Abraham Pais once told me an anecdote about Einstein and Poincaré. He had the occasion to discuss Poincaré with Einstein. In 1905, Poincaré published an important paper on Lorentz's version of the theory of relativity. This paper appeared before Einstein's. One of the things that Poincaré did in this paper was to show that the Lorentz transformations form what mathematicians call a "group." Two successive Lorentz transformations are equivalent to a third. There is an inverse to any Lorentz transformation—that sort of thing. This group is stilled called the Poincaré group. It turned out that Einstein had never read this paper. Pais lent him his copy. There was apparently no comment when Einstein returned it. He certainly knew its contents. Michelson was another case. He and Einstein met only once, in 1931. Michelson was then seventy-nine years old. On this occasion he expressed regret that his work had given birth to this "monster"– relativity. Einstein, it seems, refrained from telling him that his work had little or nothing to do with the creation of relativity.

Whatever immediate improvement there was in Einstein's material circumstances had nothing to do with his 1905 papers. It had to do with the excellence of the work he was doing in the patent office. In 1906, he was promoted to technical expert second class and his salary raised to 4,500 Swiss francs a year. Einstein's first academic job was in 1908, when he became, in addition to a patent examiner, a *Privatdozent* at the University of Bern. From what I have read, he was not a very good lecturer at this level. He just had too many original ideas in his mind to focus on teaching not very advanced students. It was in 1909, when Einstein left the patent office to become an associate professor at the University of Zurich with the same salary he had had at the patent office. From that time on he made a meteoric rise in academia. In 1911, he became a full professor at the German University in Prague. This was an official government position and involved a swearing-in ceremony in which a special uniform had to be worn. It seems as if Einstein wore it once. When Professor Frank succeeded him the following year–Einstein went to the ETH–he gave Professor Frank the uniform. Considerable alterations must have been required considering the fact that Professor Frank was substantially smaller.

Figure E.1. The attendees at the Solvay Congress of 1911. (Photographie Benjamin Couprie, Institut International de physique Solvays, courtesy AIP Emilio Segré Archives)

Einstein had barely unpacked in Zurich when he was on the move again–this time to Berlin. There were several reasons why he made this change. His job at the ETH required a considerable teaching commitment. Einstein was now deep into the research that led to the general relativity and did not want the distraction of teaching. The people in Berlin, led by Planck, made him an offer he could not refuse. It was a research position under the aegis of the Prussian Academy of Sciences with no teaching requirement, which came with a professorship at the University of Berlin. A new institute for theoretical physics was going to be created for Einstein. In addition to this, he had already started a relationship with his cousin Elsa which would lead to their marriage in 1919. It has also to be said that the scientists assembled in Berlin were the most distinguished group anywhere in the world in all fields. There was a physics colloquium that Einstein participated in which was internationally famous. He chose to teach a seminar in statistical mechanics. I knew a few people who had taken it–people like the late Eugene Wigner–who recalled it as the high point of their scientific education. He also had a small number of PhD students.

In 1919, word of the eclipse experiments that seemed to confirm Einstein's theory of gravitation, was disseminated. Einstein suddenly found himself internationally famous. It went way beyond anything anyone had known since the days of Newton. Unlike Newton, Einstein was a colorful individual who readily expressed himself on a variety of subjects. The image we have of Einstein is largely from his later years in Princeton, where he was a rather reclusive and somewhat unkempt figure. The one time I saw him, the year before he died, he was getting into a station wagon belonging to the Institute for Advanced Study. He was wearing what looked like a navy surplus pea jacket, possibly with no socks. But during his earlier years he seemed to enjoy at least some of his celebrity. Here are two examples. There was in Berlin a fashionable doctor named Janos Plesch. Plesch, who had been born in Hungary, had put together a practice that included people like Marlene Dietrich. The practice had made him rich, in addition to which he had married a very wealthy woman. He lived in a mansion in Berlin where he gave

"Herrenabends"–stag-night dinners with fabulous menus and guests who were at the forefront of Berlin cultural life. Einstein became a regular guest at these dinners. He soon became a patient. He was suffering from pericarditus–an inflammation of the membranous sac that encloses the heart. The fact that Einstein chose Plesch as his doctor scandalized many of his Berlin acquaintances who regarded Plesch as an opportunistic quack. Nonetheless, Plesch managed to cure Einstein and they remained close friends.

The second example involves Einstein's second visit to the United States. From December of 1930 until March of 1931, he spent most of his time at the California Institute of Technology. During that visit he got to know Charlie Chaplin, who invited him to the premier of his film *City Lights*. The photograph of Einstein and his wife, in elegant formal clothes, taken with Charlie Chaplin, shows, in my view, a man who was thoroughly enjoying where his celebrity had gotten him (Figure E.2). In 1932, Einstein was back once again at CalTech. It was becoming clear that his situation in Germany had become impossible. There were some rather confused negotiations as to whether Einstein would take a permanent position at CalTech. I have often speculated as to what his later career might have been like if he had. Robert Oppenheimer was in the process of developing what became the preeminent school of theoretical physics in this country. There were very active experimenters. For example, Ernest Lawrence was inventing the cyclotron and, in 1932,

Figure E.2. Einstein, Charlie Chaplin, and Elsa Einstein at the premier of *City Lights* in January 1931. (Courtesy Jewish Chronicle Limited/HIP/The Image Works)

Carl Anderson observed the first antiparticle–the positron. If Einstein had gone to CalTech would he have stayed more connected with the advancement of physics? Would the presence of these young people have stimulated him? We will never know, because at this time Abraham Flexner, who was an educator who had a mandate to create a new kind of research institution, what became the Institute for Advanced Study in Princeton. Einstein would have no duties except to do his own work. In 1932, he accepted this offer and from 1933, until his death in 1955, Princeton became his home.

Einstein continued to do his own sort of physics until the day of his death. His old friends such as Max Born tried to persuade him to accept the quantum theory and return to the mainstream. Einstein's response was that he felt that he had earned the right to make his own mistakes. I once asked Miss Dukas how she would describe Einstein's last years. The word that she used was "serene" (Figure E.3). He was very concerned

Figure E.3. Einstein in his later years. (Library of Congress, courtesy Emilio Segré Archives)

about the fate of mankind but little concerned about his own fate. What always strikes me is how we are still in the thrall of his ideas. I will give one final example that does not come from 1905, but rather from 1917. Einstein decided to apply his new theory of gravitation to a study of the universe at large—cosmology. At the time it was widely felt that the universe was just the Milky Way galaxy alone. Einstein was concerned as to what kept this universe stable—what kept it from collapsing under the effects of gravitation. He decided that the only way to avoid this was to add a new term to his equations of general relativity—a term that he called the "cosmological member." This term acted like a repulsive force that counteracted gravity. In the next decade, especially with the work of the astronomer Edwin Hubble, it became clear that the universe was in fact expanding. There seemed to be no need for the cosmological term and Einstein abandoned it. This is where things stood until a few years ago when it became clear that the universe is now expanding at a greater rate that can be accounted for by the conventional theory. The cosmological term is back and is referred to as "dark energy." With the dark energy, the Old One has once again played a trick on us, something that Einstein might have found wonderfully amusing. It seems as if the Old One has no end of secrets.

Notes

◦ Introduction

"the number is still growing." Among them here is a personal list. For people with the needed technical background *Subtle is the Lord: The Science and the Life of Albert Einstein*, Oxford University Press, New York, 1981, by Abraham Pais is the best account of the science. For a general biography, I recommend *Albert Einstein: A Biography* by Albrecht Fölsing, translated by Ewald Osers, Viking, New York, 1997. For an older biography there is *Einstein: His Life and Times*, by Philipp Frank, DaCapo Press, New York, 1953. Professor Frank knew Einstein for much of their professional lives. For a shorter biographical sketch there is my *Albert Einstein and the Frontiers of Physics*, Oxford University Press, New York, 1996. For a sensitive view of Einstein's romantic life there is *Einstein in Love* by Dennis Overbye, Penguin, New York, 2001.

"written in his honor." *Albert Einstein: Philosopher-Scientist*, edited by P. A. Schilpp, MFJ Books, New York, 1970.

"be behind things." Schilpp, p. 9.

"papers were written." A reader who wants to see translations of these papers, including the thesis, will find *Einstein's Miraculous Year*, edited by John Stachel, Princeton University Press, Princeton, 1998, very useful. Stachel's introductory essays are excellent, but require considerable technical background to follow. This is not a book for the general reader.

⤺ Chapter 1

"a month, a year." This quotation is from Newton's *Principia*, but I have taken the translation from the Latin from *The Science of Mechanics* by Ernst Mach, Open Court Publishing, Lasalle, 1974, p. 272. I will refer to this as "Mach."

"Two Chief World Systems." The edition I use, with a forward by Einstein, is published by the Modern Library, New York, 1966. I will refer to this as "Galileo."

"rapidly westward." Galileo, p. 153.

"standing still." Galileo, pp. 216–217.

"bit about him." This information and the quotations are taken from my essay, "Ernst Mach and the Quarks" in the collection *Cranks, Quarks, and the Cosmos*, Basic Books, New York, 1993, pp. 28–37.

"in these matters." Bernstein 1993, p. 30.

"harmful vermin." This quote can be found in Gerald Holton, "Mach, Einstein and the Search for Reality," *Daedalus*, Spring 1968, p. 657.

"hereafter be shown." Mach, p. 298.

"indefensible conceptions." Mach, p. 265.

"matter conjointly." Mach, p. 298.

"is impressed." Mach, p. 301.

"actual facts." Mach, p. 272.

"metaphysical conception." Mach p. 273. The italics are in Mach.

"can be referred." Mach, p. 277. The italics are in Mach.

"of the [absolute] motion." Mach, p. 277

"our imagination." Mach, p. 284.

"space is not empty." As one might imagine the issues raised here are more complex that I have indicated. The references to Thirring's papers are *Phys. ZS.* **19**, 33(1918), **22**, 29 (1921). Arguments have been made that Thirring's example is too abstract, and that even if you grant his point, it does not prove that the acceleration of rotation is not absolute. The discussion of this would take us way beyond the intent of this book. An interested reader might consult, "On Mach's critique of Newton and Copernicus," by H. Hartman and C. Nissam-Sabat, *American Journal of Physics* **71**, November 2003, pp. 1163–69.

"grove of trees." This is quoted in Volume 1 of *A History of Aether and Electricity*, by Sir Edmund Whittaker, Dover Publications, New York, 1989, pp. 108–109. I will refer to these volumes as "Whittaker 1,2."

"form or vigour." Quoted in Whittaker 1, p. 19.

"*Lorentz transformation.*" *The Feynman Lectures on Physics*, by Richard Feynman, Addison-Wesley, Reading, Massachusetts, 1975, Vol. I, pp. 15–3.

"not born in this country." Michelson's youngest daughter, Dorothy Michelson Livingston, wrote a very nice biography of her father–*The Master of Light*, Charles Scribner, New York, 1973.

"through the ether." Livingston, p. 82

"of the solar system" Pais, p. 122.

"l times smaller." This can be found in the collection *The Principle of Relativity*, by Einstein et al, Dover Press, New York, 1952, p. 21. There is also an earlier paper in the collection dealing with the Michelson-Morely experiment. I will refer to this collection as "Einstein et al."

"but *rigorously*." Whittaker 2, p. 30.

"*relative* displacements." Whittaker 2, p. 30. Italics here and in the following quotation are in Whittaker.

"*velocity of light*." Whittaker 2, pp. 30–31. Poincaré does not explain how he arrived that this conclusion about the speed of light being the maximum possible speed, but it is not difficult to see how he might have. If you assume that the Lorentz-Fitzgerald contraction is given for all speeds by the factor $\sqrt{1 - \frac{v^2}{c^2}}$ it follows that when $v = c$ the object is contracted to zero. Moreover, if v becomes greater than c, then we have the square root of a negative number which cannot be interpreted as a length.

↜ CHAPTER 2

"much attention." Whittaker 2, p. 40.

"Einstein lecture." In 1921, at the age of twenty-one, Pauli published a masterful monograph on relativity. It is still one of the best. The English translation is *Relativity Theory*, Dover, New York, 1981.

"from the essential." Schilpp, p. 15.

"Method of Theoretical Physics." The lecture was apparently originally written in German. The pamphlet Einstein sent me was in English. It can be found in *Essays in Science* by Albert Einstein, Philosophical Library, New York, 1934. The quote from it can be found on page 17.

"acute angles." Schilpp, pp. 10–11.

"independent scientific work." Schilpp, p. 15.

"simultaneous events." The quote is from Einstein's 1905 relativity paper reprinted in Einstein et al, p. 39.

"philosophical writings." Schilpp, p. 53.

"*A Popular Exposition*." Methuen, London, 1920.

"valuable suggestions." Einstein et al, p. 65.

"as *results*." This quotation can be found in my *Quantum Profiles*, Princeton University Press, Princeton, New Jersey, 1991 in an essay entitled "Besso," which has details about his life. The quotation is on p. 150.

"special properties." Einstein et al, pp. 37–38.

"one readily shows." See, for example, *Modern Physics*, by Jeremy Bernstein, Paul Fishbane and Stephen Gasiorowicz, Prentice Hall, New Jersey, 2000, pp. 44–45. In my opinion the best textbook on relativity is *Relativity: Special,General and Cosmological* by Wolfgang Rindler, Oxford University Press, New York, 2001.

"have in mind" For a very clear discussion of this and related matters at about the same level as what I am trying to achieve see http:/www.phys.virginia.edu/classes/252/srelwhat.html. This is a site created by Professor Michael Fowler of the University of Virginia.

"go into here." See, for example. Bernstein et al, pp. 54–55.

"stellar aberration." A nice treatment of this at about the same level can be found in "Einstein's Special Theory of Relativity and the Problems in the of Electrodynamics of Moving Bodies that Led him to it." By John D. Norton, *Cambridge Companion to Einstein*, edited by M. Janssen and C. Lehner, Cambridge University Press, New York, 2005.

"innumerable places." See, for example, Bernstein et al or Rindler, or Einstein et al for the original derivation. For a reader with some technical background *Albert Einstein's Special Theory of Relativity* by Arthur I. Miller, Addison Wesley, 1981, is highly recommended both for its science and history.

"this result is confirmed." For details see Bernstein et al Chapters 2 and 17. The experiments were done by a team under the direction of Carroll O. Alley. Professor Alley has summarized the work in "Proper Time Experiments in Gravitational Fields with Atomic Clocks, Aircraft and Laser Light Pulses" in *Quantum Optics, Experimental Gravity and Measurement Theory*, edited by Pierre Meystre and Marlon O.Scully, Plenum, New York, 1983, pp. 363–427.

"in the loop." For a very clear discussion see Miller, Chapter 3.

"successfully put to the test." Einstein et al, p. 71.

"another man's past." This is quoted in *Understanding Relativity*, by Stanley Goldberg, Birkhäuser, Boston, 1984, p. 261. Magie took the odd position that relativity could not be a fundamental theory since, he argued, that a fundamental theory must be understandable to everyone.

"and common sense" *Time and Free Will: An Essay on the Immediate Data of Consciousness*, translated by F.L. Pogson, George Allen and Unwin, London, 1910, p. 2227

"end of his Latin." The original of this extraordinary letter, which was written in March of 1906 and is in French, can be found in Miller, pp. 336–337.

"complexes of phenomena." This is quoted in Fölsing, p. 206.

"in Cologne." A translation of this lecture can be found in Einstein et al, pp. 75–91.

"independent reality." Einstein et al, p. 75.

"learned arguments." See, for example, Miller, pp. 257–274.

"theory of gravitation." Pais, p. 179.

"somewhere in space." A description of this can be found in Einstein's popular book, *Relativity: The Special and General Theory*, Crown Publishers, New York, 1961, Chapter XX.

"Propagation of Light." See Einstein et al p. 99 et seq. This paper is an adumbration of a paper he wrote in 1907.

"General Relativity." Einstein et al, p. 111.

"is correct." This is quoted, for example in Bernstein 1982, p. 144.

~ Chapter 3

"so-called atoms." This is quoted in *Molecular Reality* by Mary Joe Nye, American Elsiever, New York, 1972, pp. 4–5. Kekulé was one of the greatest of the 19th-century organic chemists.

"construction of matter." Schilpp, p. 19

"our sight." An excerpt from the poem can be found in the very useful two-volume *The World of the Atom*, edited by Henry A. Boorse and Lloyd Motz, Basic Books, New York, 1966. This quote is found in Vol. 1, p. 17.

"First Creation." Boorse, Vol. 1, p. 102.

"very rapid motion." Boorse, Vol. 1, pp. 112–116.

"cohesion." This work is described in articles by G.D.Scott and I.G. MacDonald, *American Journal of Physics* 33, 163, 1965 and A.P. French, *American Journal of Physics* 35, 162, 1967. See also *The Kind of Motion We Call Heat*, by Stephen Brush, North-Holland, New York, 1976, p. 75

"*Journal de Physique.*" An English translation of this paper can be found on the Web site http://web.lemoyne.edu/~giunta/avogadro/html.

"Jan Josef Loschmidt" A useful Web site is www.loschmidt.cz/biography.html. The great 1865 paper can be found in translation on www.dbhs.wvusd.k12. ca.us/webdocs/Chem-History/Loschmidt-1865.html.

"as follows." For a more detailed version of what Loschmidt did, including the factors, see *The Dictionary of Scientific Biography*, edited by Charles Gillespie, Scribners, New York, 1973, Vol. VIII, pp. 505–511. The situation is complicated by the fact that air is a mixture if gasses of which oxygen and nitrogen are the most prominent. Loschmidt took this into account.

"in my head." Bernstein, 1993, p. 34.

"judgement in the matter." This is a translation taken from Einstein's paper *Die von der molekularkinetischen Theorie der Wärme geforderte Bewegung von in ruhenden*

Flügiskeiten suspendierten Teilchen published in *Annalen der Physik,* Ser. 4, vol. **17,** 1905, pp. 549–560. The translation can be found in *Albert Einstein Investigations on the Theory of Brownian Movement,* edited by R. Fürth and translated by A.D. Cowper, Dover Publications, New York, 1956, p. 1.

"of a substance." Fürth, p. 76.

"of Heat." Fürth, p. 18.

"precision experiments." For details of this and other aspects of Perrin's life and career, see Nye.

"the present day." Nye, p. 168.

⤳ CHAPTER 4

"the quantum theory." Frank, p. 98.

"for free." This advertisement and other information about the "Olympia Academy" can be found on www.einstein-website.de/z_biography/olympia. html.

"interest you." This translation I have taken from Fölsing, p. 120.

"*maximum zu.*" Excepts in English from Clausius's papers on entropy can be found at http://web.lemoyne.edu/-giunta/Clausius.html.

"nature of the walls." Schilpp, pp. 37–38.

"Max Planck." This is quoted in "Thermodynamics and Quanta in Planck's Work," by Martin J. Klein, *Physics Today,* **19,** No. 11, 1966, p. 24.

"race." Fölsing, p. 345.

"laws of thermodynamics." This is taken from *Black Body Theory and the Quantum Discontinuity, 1894–1912* by Thomas S. Kuhn, University of Chicago Press, Chicago, 1978, p. 14.

"continuous matter." Kuhn, p. 23.

"non-human ones." Kuhn, p. 26.

"inclined to suspect." Kuhn, p. 31.

"that it is correct." See for example Kuhn.

"in space." This quotation is taken from Einstein's 1905 paper. The German reference is *Über einen die Erzeugung und Verwandlung des Lichtes betreffenden heuristischen Gesichtspunkt, Annlen der Physik,4*, Vol. **17**, 1905, pp. 132–148. I have used the translation that can be found in Boorse, Vol.1, p. 545.

"taking a risk." Pais, p. 382.

"guide investigation" This definition and more about the word can be found at http://www.websters-online dictionary.org/definition/English/he/heuristic .html.

"same law." Stachel, p. 150.

"only as units." Boorse, Vol. 1, p. 545.

Epilogue-afterword

"As to the ether" Bernstein, p. 105.

"Einstein is one " Bernstein, p. 103.

Index

Index

Index

Index